T0313369

# Ultra-Low Input Power Conversion Circuits based on Tunnel-FETs

# RIVER PUBLISHERS SERIES IN CIRCUITS AND SYSTEMS

*Series Editors*

**MASSIMO ALIOTO**
*National University of Singapore*
*Singapore*

**KOFI MAKINWA**
*Delft University of Technology*
*The Netherlands*

**DENNIS SYLVESTER**
*University of Michigan*
*USA*

Indexing: All books published in this series are submitted to the Web of Science Book Citation Index (BkCI), to CrossRef and to Google Scholar.

The "River Publishers Series in Circuits & Systems" is a series of comprehensive academic and professional books which focus on theory and applications of Circuit and Systems. This includes analog and digital integrated circuits, memory technologies, system-on-chip and processor design. The series also includes books on electronic design automation and design methodology, as well as computer aided design tools.

Books published in the series include research monographs, edited volumes, handbooks and textbooks. The books provide professionals, researchers, educators, and advanced students in the field with an invaluable insight into the latest research and developments.

Topics covered in the series include, but are by no means restricted to the following:

- Analog Integrated Circuits
- Digital Integrated Circuits
- Data Converters
- Processor Architectures
- System-on-Chip
- Memory Design
- Electronic Design Automation

For a list of other books in this series, visit www.riverpublishers.com

# Ultra-Low Input Power Conversion Circuits based on Tunnel-FETs

### David Cavalheiro

Universitat Politècnica de Catalunya
Spain

### Francesc Moll

Universitat Politècnica de Catalunya
Spain

### Stanimir Valtchev

Universidade Nova de Lisboa
Portugal

Routledge
Taylor & Francis Group

LONDON AND NEW YORK

**Published 2018 by River Publishers**
River Publishers
Alsbjergvej 10, 9260 Gistrup, Denmark
www.riverpublishers.com

**Distributed exclusively by Routledge**
4 Park Square, Milton Park, Abingdon, Oxon OX14 4RN
605 Third Avenue, New York, NY 10017, USA

*Ultra-Low Input Power Conversion Circuits based on Tunnel-FETs* / by
David Cavalheiro, Francesc Moll, Stanimir Valtchev.

Routledge is an imprint of the Taylor & Francis Group, an informa
business

ISBN 978-87-93609-76-1 (print)

While every effort is made to provide dependable information, the
publisher, authors, and editors cannot be held responsible for any errors
or omissions.

*To our families*

# Contents

# Preface

There has been a tremendous evolution in integrated circuit technology in the past decades. With the scaling of complementary metal-oxide-semiconductor (CMOS) transistors, faster, less power consuming, and more complex chips per unit area have made it possible for electronic gadgets to evolve to what we see today.

The increasing demand in electronic portability imposes low power consumption as a key metric to analog and digital circuit design. While dynamic power consumption decreases quadratically with the decrease of power-supply voltage, leakage power presents a limitation due to the inverse sub-threshold slope (SS) of conventional CMOS devices. A power supply reduction implies a consequent threshold voltage reduction that, given the fixed SS, causes an exponential increase in leakage current. This poses a limitation in the reduction of power consumption that is inherent to the conventional thermionic-based transistors (MOSFETS and FinFETs). In thermionic emission-based transistors the SS at room temperature is limited to 60 mV/dec.

To circumvent the SS limitation of conventional transistors, devices with different carrier-injection mechanisms are required. The Tunnel Field-Effect Transistor (TFET) is presented as the most promising post CMOS technology due to its non-thermal carrier-injection mechanism based on Band-To-Band Tunneling (BTBT) effect. TFETs are known as steep slope devices (SS $<$ 60 mV/dec at room temperature). Large current gain ($I_{ON}/I_{OFF} > 10^5$) at low voltage operation (sub-0.25 V) and extremely low leakage current have already been demonstrated, placing TFETs as serious candidates for ultra-low-power and energy-efficient circuit applications. TFETs have been explored mostly in digital circuits and applications.

In this book, the use of TFETs is explored as an alternative technology also for ultra-low-power and voltage conversion and management circuits, suited for weak energy harvesting (EH) sources. As TFETs are designed as reverse-biased $p$-$i$-$n$ diodes (different doping types in source/drain regions), the particular electrical characteristics under reverse bias conditions require changes in conventional circuit topologies. In this book, ultra-low input-power conversion circuits based on TFETs are designed and analyzed, evaluating their performance with the proposal of rectifiers, charge pumps, and power-management circuits (PMC). TFET-based PMCs for RF and DC EH sources are proposed and limitations (with solutions) of using TFETs in conventional inductor-based boost converters identified.

# Acknowledgments

The authors wish to acknowledge the financial support given by the Portuguese funding institution FCT (Fundação para a Ciência e a Tecnologia), the Spanish Ministry of Economy (MINECO), and ERDF funds through project TEC2013-45638-C3-2-R (Maragda).

# List of Figures

# List of Tables

# 1

---

# Introduction

---

## 1.1 The Technology Scaling Roadmap so far

Since 1947, with the invention of the first transistor by William Shockley, John Bardeen, and Walter Brattain at Bell Labs [1] and the Integrated Circuit (IC) at Texas Instruments by Jack Kilby in 1958 [2], the impact of the evolution of electronic technology in our daily activities has been so enormous that nowadays it is unthinkable to live without it. Smart cars, smart phones, smart watches, smart TVs, and healthcare gadgets are just a few examples of technology that eases our daily life, mainly due to the downscale evolution of the electronic transistor. The exponential growth of the transistor count on a die, following Moore's law [3] has been the major impulse for the semiconductor industry over the years. The decrease of the technology node and the consequent transistor channel length has led to the possibility of adding more devices on a single die, thus reducing the production cost of a chip and increasing its complexity. Also, the reduction of the transistor size allowed the design of faster circuits with reduced power consumption (per transistor).

Until the late 1990s, the transistor scaling theory of Dennard's [4] was well followed by the semiconductor industry, i.e., the power-supply voltage $V_{DD}$ and threshold voltage $V_{TH}$ of the transistor decreased linearly with the reduction of the channel length and width dimensions. Consequently, with the decrease of $V_{DD}$, a quadratic decrease of the dynamic power consumption in the transistor and hence in the IC was observed over the years. The successful downscaling of the transistors was mainly achieved due to the excellent material and electrical properties of $SiO_2$, the material used in the dielectric between the gate and the channel of the device.

During the past 15 years, several modifications in the transistor structure were required in order to maintain the technology downscaling trend. As an example, strained silicon technology was introduced in the 90-nm technology node in order to improve the carrier mobility inside the transistor. This was done by some approaches such as embedding SiGe materials in the P-type metal-oxide-semiconductor field effect transistor (MOSFET) (PMOS) source/drain regions and tensile silicon nitride-capping layer for N-type MOSFET (NMOS) devices [5]. With reduced gate oxide thickness, the increase of gate leakage due to current flowing through the thin gate insulator (by tunneling) was presented as the major problem to be solved at 45 nm technology nodes. The introduction of gate dielectrics with a large dielectric constant ($\varepsilon_r \approx 25$ for $HfO_2$ compared to that of $SiO_2$, $\varepsilon \approx 3.9$) significantly reduced the gate leakage, thus allowing the decrease of the size of the technology node down to 32 nm [6]. With further reduction of channel length, the increase of thermal diffusion of carriers, and the consequent increase of leakage current required changes in the conventional bulk-complementary metal-oxide semiconductor (CMOS) structure. To overcome the consequent increase of static power consumption in chips, different device structures such as Fully Depleted Silicon on Insulator (FDSOI) and FinFETs were developed and are currently under production with technology nodes as low as 7 nm [7].

FinFETs are known as multi-gate devices. They are characterized by a gate electrode wrapped around several sides of the conducting channel, replacing the planar configuration of the conventional single-gate MOSFETs. Transistors with multi-gate configuration increase the electrostatic control of the gate over the channel, thus allowing the reduction of short-channel effects and the consequent reduction of leakage power inside the device [8].

FDSOI devices are characterized by an ultra-thin layer of insulator placed on top of the silicon base and below a non-doped thin silicon-based channel. Despite the superior electrostatic control of the gate over the channel compared to bulk-CMOS, FDSOI allows the modulation of the threshold voltage $V_{TH}$ of the device by changing the polarity of the body bias [9].

As shown in Figure 1.1, during the next few years, transistors are expected to be reduced to a few nanometers, and further miniaturization of the transistor will be practically impossible [10]. As the dimensions of MOSFETs are approaching a scale at which they will be composed of just a few hundred atoms, undesirable effects such as gate tunneling will prevent further

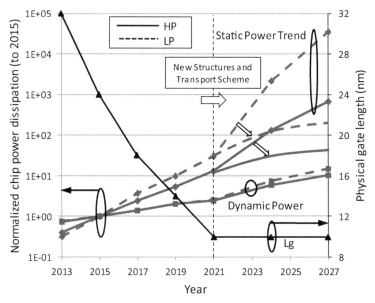

**Figure 1.1** Static and dynamic power dissipation trend in function of technology gate length.
*Source*: ITRS 2015 [11].

improvements in device performance due to large leakage currents. In order to avoid the consequent increase of the static power trend of chips with further technology downscale and keep the increase of transistor density with a viable economical production, the semiconductor industry will eventually push efforts in the development of vertical device geometries, circuitry with multiple layers (3D integration), and different electrical transport schemes [11]. In order to accomplish this, the development of devices with alternative materials, namely SiGe, Ge, and compounds drawn from groups III–V of the periodical table is mandatory. To summarize, Figure 1.2 presents the CMOS technology scaling roadmap so far and emerging devices under research for further miniaturization of transistors.

## 1.2 New Solutions for Future Technology Nodes

In order to face up to the increase of static power consumption trend in chips for future technology nodes, changes in the channel material/structure are required [11]. The most promising technologies to keep the downscaling technology to a few nanometers are listed below:

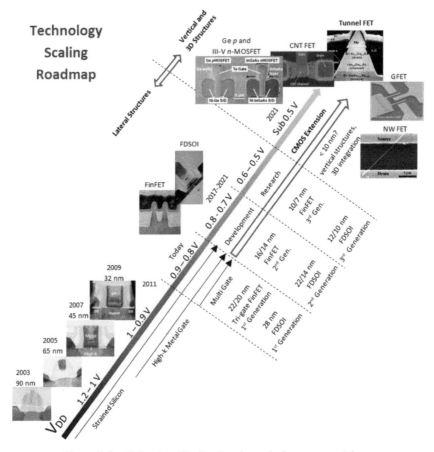

**Figure 1.2**   Technology Scaling Roadmap: before, now, and then.

- Ge and III–V materials: According to the ITRS 2015 previsions, the next step in the scaling roadmap will rely on the replacement of the strained silicon channel of MOSFETs by high mobility materials [11]. This task is challenging as many factors still have to be overcome; improvement of high-K/*Ge*–III–V interface, minimization of Band-to-Band Tunneling (BTBT) in narrow bandgap channel materials, and very-large-scale integration (VLSI) using a manufacturable process flow on a silicon platform are presented as some examples [12, 13].
- Nanowires (NWs): The replacement of conventional planar MOSFET channels with NWs may allow further decrease of the technology node due to the possibility to manufacture NWs with diameters of a few

nanometers [14]. NWs match well with gate-all-around structures that may enable the reduction of short-channel effects. They are grown vertically for the benefit of future vertical integration of devices in chips. In order to grow defect-free NWs, device yield and uniformity has to be improved, as well as position registry if the transferring of NWs to a different substrate is required. In order to avoid surface roughness and defects, proper surface treatment and passivation techniques need to be developed.

- Carbon Nanotubes (CNT): With ultra-thin body diameters as low as $\approx 1$ nm, CNT-FETs are presented as a viable option for sub-10 nm technology nodes. Compared to silicon, CNTs can offer improved electron and hole mobility in the channel at room temperature. Purified and sorted CNTs with relatively uniform diameter distribution are presented as some problems that have to be overcome for VLSI integration [15].

- Graphene Field-Effect Transistors (GFETs): With higher carrier mobility than CNTs, graphene materials can benefit the RF design community by allowing faster transistors with reduced dimensions. As graphene is presented as a single atomic layer, it is presented as a zero bandgap semiconductor. Therefore, the main research relies on opening a bandgap in order to efficiently allow a large current gain ($I_{ON}/I_{OFF}$) for both analog and RF applications [16].

- Tunnel FETs (TFETs): The TFET is considered the most promising switching technology for low-power, low-performance (LP) applications due to its unique electrical characteristics [17]. With a different carrier injection mechanism than conventional thermal devices, TFETs can achieve an inverse sub-threshold slope (SS) with sub-60 mV/dec at room temperature. With this characteristic, TFETs have the potential to achieve a low operating voltage, maintaining a large current gain. The future integration of TFETs in low power chips is strongly dependent on the evolution of III–V manufacturing processes as TFETs designed with groups III and V materials overcome the electrical performance of Si-based TFETs [18].

Further improvements in conventional transistor architectures and the integration of new materials may not only maintain the downscaling of the technology roadmap but also enhance the functionality and performance of future electronic systems. However, as the dimensional scaling of current technology will eventually approach fundamental limits, different trends than the current "More Moore" (MM) will emerge.

## 1.3 Energy Harvesting in a More than Moore era

Nowadays, the industry is pursuing a new trend denominated "More than Moore" (MtM), where the improved performance of new technologies is not only traded against power, but also against functional diversification of semiconductor-based devices [11]. The rise of new materials and emerging technologies can open doors to further improvement in areas that do not necessarily scale at the same rate as that of digital functionalities [e.g., sensors, actuators, biochips, RF, analog design, energy harvesting, power management, Internet of Things (IoT), etc.]. Therefore, the heterojunction integration of MM (digital) and MtM (non-digital) functionalities into compact integrated systems is expected to further improve the performance of a wide variety of applications such as in communications, healthcare, security, and automobiles, among others, where performance is not the main metric to accomplish but rather functionality.

The market IoT is one that will certainly benefit from the integration of MM and MtM trends. The possibility of wireless interconnection of any device through the Internet or local networks can enable objects to be sensed/controlled remotely, creating opportunities in several areas where systems have to perform actions or sense the surrounding environment. The IoT market is expected to create a huge network of billions or trillions of devices communicating with one another and therefore, for reasons of cost, availability, and convenience, one of the major challenges for this system integration is the replacement of the battery by green solutions such as energy harvesting (EH) [19].

In Figure 1.3, a possible architecture of a self-powered sensor node, comprising MM technologies for digital processing, MtM for sensing and power management units, and energy harvesting for powering the whole system is considered. In order to power an integrated system such as the one presented, or any other system with energy from the surrounding environment, several challenges have to be solved: first, the digital processing and storage unit have to be extremely energy efficient, requiring the lowest possible power consumption for proper operation. This will minimize the energy required for proper system functioning. Second, the power management circuit (PMC) has to consume less power than that produced by the energy harvesting transducer. This will enable the storage of energy in a charge tank to be further used as a power supply source for the entire system.

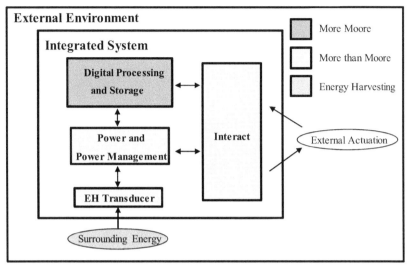

**Figure 1.3** Integration of More Moore, More than Moore, and Energy Harvesting for IoT sensors.

## 1.4 Tunnel FETs as a Key Technology for Energy Harvesting

A solution to minimize the power consumption of the digital processing unit is to decrease the power-supply voltage $V_{DD}$ of the digital circuitry. A linear decrease of $V_{DD}$ will result in a quadratic decrease of the dynamic power consumption of the digital circuitry. However, if the threshold voltage $V_{TH}$ of the transistor does not scale proportionally with $V_{DD}$, the leakage current and consequent static power consumption of the system will suffer from an exponentially increase. This is directly related to the thermal-dependent carrier-injection mechanism of conventional MOSFETs: the inverse SS is limited by 60 mV/dec (at room temperature), and therefore a decrease in $V_{DD}$ results in an exponential increase of sub-threshold leakage current according to the equation expressed by (1.1) [20]:

$$I_{sub} = K_1 W e^{-\frac{V_{TH}}{nV_\Theta}} \left( 1 - e^{-\frac{V_{DD}}{V_\Theta}} \right) \tag{1.1}$$

In the sub-threshold leakage current expression, $K_1$ and $n$ are constants experimentally derived, $W$ is the width of the transistor and $V_\Theta$ is the thermal voltage with a value of 25 mV at room temperature. As shown in Figure 1.4,

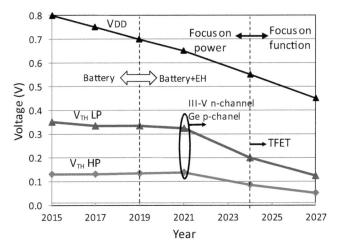

**Figure 1.4**    Voltage trend of logic technology.

*Source:* ITRS 2015 [11].

the threshold voltage of transistors for both high-performance (HP) and LP applications are expected to be kept constant during the next few years [11]. With the decrease of the power supply voltage $V_{DD}$, an increase of leakage power in chips will be expected as shown by the power trends shown in Figure 1.1. With the introduction of different technologies with new materials and transport schemes (not thermal-dependent), the reduction of the threshold voltage in transistors will be possible without the cost of increased static/leakage power consumption of chips.

There is currently a great deal of research on switches with steep SS, i.e., sub-60 mV/dec (at room temperature). Among several options and with a different carrier-injection mechanism, the TFET device is presented as the most promising switch technology for low-voltage operation (sub-0.5 V) and low performance applications [21, 22]. As presented in Figure 1.5, the TFET device allows for a decrease of the overdrive voltage ($V_{GS}-V_{TH}$) maintaining superior performance in terms of leakage current and the consequent static power consumption when compared to conventional thermal MOSFETs. With improved performance at low voltage, the TFET device is presented as a natural candidate to be applied to ultra-low power, LP ICs (e.g., IoT sensor nodes) powered by the surrounding energy [11].

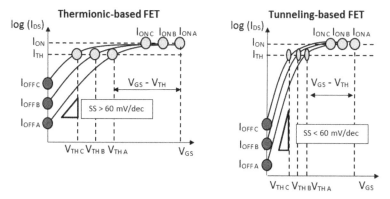

**Figure 1.5** MOSFET and Tunnel FET input characteristics.

## 1.5 Topics Addressed in This Book

With the increased demand for electronic portability, the chip-industry market is forcing "ultra-low power" as a key metric to achieve in order to extend the battery life of future electronic gadgets, or, if possible, to use the surrounding energy as the main power source. However, this metric contrasts with the trend shown by recent technologies for increase of power consumption at reduced dimensions. In order to avoid the increase of static power that degrades the performance of ultra-low power circuits with reduced technology nodes, technologies with different carrier injection mechanisms are required. The TFET device is shown in the technology-scaling roadmap as a candidate for future technology nodes. It can allow the reduction of both static and dynamic power consumption in circuits due to the improved electrical characteristics at low-voltage operation. In the literature, most of the works based on TFETs are related to logic design and techniques to improve the device performance. There are few works exploring the performance of TFETs in analog applications, and a clear lack of works investigating the performance of this technology in PMC for energy harvesting applications.

Considering the above-mentioned points, the work presented in this book aims to investigate the electrical performance of TFET devices, exploring the opportunities and drawbacks of using this emerging technology in the design of ultra-low power circuits for energy harvesting applications. In order to accomplish this goal, this book focuses on the design of TFET-based power conversion circuits and PMC, exploring their limitations and providing circuit architectural solutions to leverage on the different TFET electrical characteristics. To accomplish the objectives, several topics are addressed as follows:

- **At a device level, evaluate the electrical properties of TFETs with the impact of physical parameters:** With the support of Atlas Device Simulator from Silvaco [23], the dependence of the TFET current-voltage characteristics on the dimensions and materials of the channel and gate dielectric has to be evaluated. This part is important as it highlights the most important physical parameters to take into account at a device level, for further improvements of TFET-based circuit performance.

- **To determine figures of merit reflecting both the digital and analog performance of TFET devices:** With an optimized TFET model for circuit simulations, a comparison between several figures of merit between TFETs and thermionic MOSFETs, at a device level, for both analog and digital design has to be performed. This task will identify the voltage range where TFETs present improved electrical performance in comparison to conventional technologies.

- **To evaluate the performance of TFET-based front-end circuits for energy harvesting applications:** As charge-pumps and rectifiers are usually the circuits interfacing the energy harvesting transducer with the power management unit, a study on the application of TFETs in such front-end circuits is required. This task will identify the voltage/power levels where the integration of TFETs is advantageous compared to the application of conventional thermionic MOSFETs. Advantages and drawbacks have to be identified, with the proposal of circuit-solutions to improve the performance of TFET-based front-end circuits.

- **Propose TFET-based PMC for Radio Frequency (RF) and Direct Current (DC) energy harvesting sources:** With a previous study on TFET-based circuits for both analog and digital design, and the proposal of front-end circuits, PMCs have to be designed considering the different electrical characteristics of this technology. Similar to the previous point, the advantages and drawbacks have to be identified, with the proposal of circuit-solutions to further improve the performance of TFET-based PMCs.

## 1.6  Book Structure

In order to accomplish the previous topics, this book was written with the framework presented in Figure 1.6. In Chapter 2, the TFET state of the art is presented. The BTBT carrier injection mechanism of TFETs is explained, and an historical review of the TFET structure is presented.

In Chapter 3, a study on the TFET current-voltage dependence on several physical parameters is performed. This task identifies key parameters for enhanced TFET-based circuit performance. In Chapter 4, optimized TFET models are simulated in order to compare the electrical characteristics of TFETs and conventional thermionic MOSFETS (such as FinFETs) for digital and analog applications.

Chapter 5 and Chapter 6 perform a study on the implementation of TFETs in front-end circuits for energy harvesting applications: charge-pumps and rectifiers. Solutions to improve the performance of such TFET-based front-end circuits are explored and compared with conventional circuit solutions

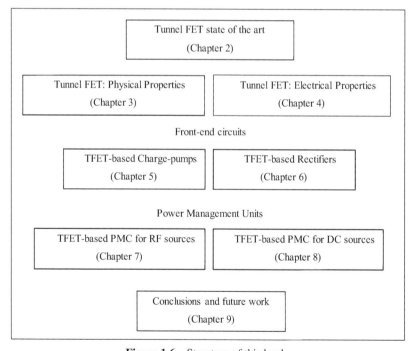

**Figure 1.6**   Structure of this book.

found in the literature. These two chapters identify the range of voltage and power levels where TFET-based converters present improved performance in comparison to that of thermionic-based counterparts.

Chapter 7 and Chapter 8 propose TFET-based PMCs for energy harvesting applications. Such circuits are designed with the previously studied TFET-based front-end circuits (rectifier and charge-pump for RF PMC and charge-pump for DC PMC). In Chapter 7, a PMC for $\mu$W RF energy harvesting applications is proposed, designed and optimized (considering the particular TFET electrical characteristics) with a startup circuit, controller, and boost converter. The limitations and advantages of using TFETs in such circuits are identified. Chapter 8 is presented as an extension of Chapter 7. With the previous limitations identified, Chapter 8 proposes solutions to enhance the performance of the PMC at decreased power levels. A study of the proposed *TFET*-based PMC powered by nW DC energy harvesting sources is presented.

Finally, Chapter 9 highlights the main conclusions of this book, discussing the results and pointing to directions for future research.

## References

[1] Shockley, W. "The path to the conception of the junction transistor," in *IEEE Transactions on Electron Devices*, vol. 23, pp. 597–620, 1976.

[2] Kilby, J. S. "Turning Potential into Realities: The Invention of the Integrated Circuit," Available at: http://nobelprize.org/nobel_prizes/physics/laureates/2000/kilby-lecture.pdf, 2000.

[3] Moore, G. E. "Cramming more components onto integrated circuits," *Reprinted from Electronics*, vol. 38, no.8, pp. 114–117, 1965.

[4] Dennard, R. H., Gaensslen, F. H., Rideout, V. L., Bassous, E., and LeBlanc, A. R. "Design of ion-implanted MOSFET's with very small physical dimensions," in *IEEE J. Solid-State Circuits*, vol. 9, pp. 256–268, 1974.

[5] Thompson, S. E., Armstrong, M., Auth, C., Alavi, M., Buehler, M., Chau, R., et al. "A 90-nm logic technology featuring strained-silicon," in *IEEE Transactions Electron Devices*, vol. 51, no. 11, pp. 1790–1797, 2004.

[6] Chau, R., Datta, S., Doczy, M., Doyle, B., Kavalieros, J., and Metz, M. "High-k/metal-gate stack and its MOSFET characteristics," in *IEEE Electron Device Letters*, vol. 25, no.6, pp. 408–410, 2004.

[7] Wu, S. Y., et al. "A 7nm CMOS platform technology featuring 4th generation FinFET transistors with a $0.027\mu m^2$ high density 6-T SRAM cell for mobile SoC applications," in *IEEE International Electron Devices Meeting (IEDM)*, pp. 2.6.1–2.6.4, 2016.

[8] Ferain, I., Colinge, C. A., and Colinge, J. P. "Multigate transistors as the future of classical metal-oxide-semiconductor field-effect transistors," *Nature* 479, 310–316, 2011.

[9] Sugii, N. "Road to $V_{min}$ = 0.4V LSIs with least-variability FDSOI and back-bias control," in *SOI Conference, 2011 IEEE International*, pp. 1–19, 2011.

[10] Courtland, R. "Transistors could stop shrinking in 2021," in *IEEE Spectrum*, vol. 53, no. 9, pp. 9–11, 2016.

[11] ITRS. Available at: "http://www.itrs2.net/," International Technology Roadmap for Semi-conductors, 2015.

[12] Nainai, A., Raghunathan, S., Witte, D., Kobayashi, M., Irisawa, T., Krishnamohan, T., et al. "Engineering of strained III–V heterostructures for high hole mobility," in *IEEE International Electron Device Meeting,* pp. 1–4, 2009.

[13] Tsipas, P., and Dimoulas, A. "Modeling of negative charged states at the Ge surface and interfaces," *Appl. Phys. Lett.* 94:012114, 2009.

[14] Luryi, S., Xu, J., and Zaslavsky, A. "Nanowires: Technology, physics and perspectives," *in Future Trends in Microelectronics: From Nanophotonics to Sensors to Energy*, 1, Wiley-IEEE Press, pp. 171–181, 2010.

[15] Kim, Y. "Integrated circuit design based on carbon nanotube Field Effect Transistor," *Transactions on Electrical and Electronic Materials*, vol. 12, no. 5, pp. 175–188, 2011.

[16] Akinwande, D., Petrone, N., and Hone, J. "Two-dimensional flexible nanoelectronics," *in Nature communications*, vol. 5, no. 5678, 2014.

[17] Ionescu, A. M., and Riel, H. "Tunnel field-effect transistors as energy-efficient electronic switches," *Nature* 479, 329–337, 2011.

[18] Avci, U. E., Morris, D. H., and Young, A. "Tunnel Field-Effect Transistors: Prospects and Challenges," in *IEEE Journal of the Electron Devices Soc.*, vol. 3, no. 3, pp. 88–95, 2015.

[19] Chen, S., Xu, H., Liu, D., Hu, B., and Wang, H. "A vision of IoT: Applications, challenges, and opportunities with china perspective," in *IEEE Internet of Things Journal*, vol. 1, no. 4, pp. 349–359, 2014.

[20] Kim, N. S., Austin, T., Baauw, D., Mudge, T., Flautner, K., Hu, J. S., et al. "Leakage current: Moore's law meets static power," in *Computer*, vol. 36, no. 12, pp. 68–75, 2003.

[21] Ionescu, A. M., De Michielis, L., Dagtekin, N., Salvatore, G., Cao, J., Rusu, A., et al. "Ultra low power: Emerging devices and their benefits for integrated circuits," *Electron Devices Meeting (IEDM), IEEE International*, pp. 16.1.1–16.1.4, 2011.

[22] Nikonov, D. E., and Young, I. A. "Uniform methodology for benchmarking beyond-CMOS logic devices," in *Technical Digest-International Electron Devices Meeting, IEDM*, pp. 25.4.1–25.4.4, 2012.

[23] Atlas User's Manual, Silvaco, Inc., Santa Clara, CA, United States, November 7, 2014.

# 2

# Tunnel FET: State of the Art

## 2.1 The Tunneling Phenomenon

Back in 1958, the Japanese scientist L. Esaki at Sony Corporation was the first to demonstrate a device working under the principle of Band-to-Band Tunneling (BTBT): the tunnel diode [1]. The principle of operation based on the laws of quantum theory was shown to be different from the transistors, ordinary diodes, and other semiconductor devices of that time. In a tunnel diode, carriers can disappear from one side of a potential barrier and appear instantaneously on the other side, even if the carrier does not have sufficient energy to surmount the barrier. It is like the carrier can "tunnel" underneath the barrier which is the space charge depletion region of the *p-n* junctions. At first, the potential of the tunnel diode was not recognized, much due to the lack of comprehension of the tunneling behavior. After many decades of investigation, the rich amount of information about tunneling processes has turned the tunneling effect into a possible solution to emerging switching devices due to the advantages shown at low-voltage operation (sub-0.5 V).

During the investigation of the internal field emission in semiconductor diodes with heavily-doped germanium (Ge) junctions, a non-monotonic current-voltage characteristic under forward bias conditions and low temperatures was observed by Esaki. The elasticity of the tunneling process using Ge materials resulted in a negative differential resistance (NDR) effect, where electron energy was shown to be conserved during the tunneling process. Due to the great importance of the tunneling effect in semiconductors, Esaki was awarded with the Nobel Prize in Physics in 1973. Radio transmitters and receptors, amplifiers, computation, and DC to AC converters were some of the first areas to benefit from the NDR effect in the current-voltage characteristic of the tunnel diode [2].

In Figure 2.1, the NDR effect in the current-voltage characteristics of a tunnel diode is presented and explained graphically. With high doped junctions and considering thermal equilibrium, the Fermi levels are located within the allowed bands as shown in Figure 2.1(b). Above the Fermi level there are no filled states (electrons) and below the Fermi level no empty states (holes) available on the regions [3]. When a differential of potential between regions occurs, electrons may tunnel from the conduction band to the valence band (and vice-versa) if some conditions are verified: occupied energy states exist on the side from which the electron tunnels; unoccupied energy states exist at the same energy level on the side to which the electron can tunnel; the tunneling potential barrier height is low and the barrier width is small enough that there is a finite tunneling transmission probability; the electron energy is conserved during the tunneling process.

As shown in Figure 2.1(c), under a forward-bias condition, energy bands exist between regions in which there are filled states in the n-region and unoccupied states in the p-region. Therefore, electrons can tunnel from one region to another. As shown in Figure 2.1(d) and with a further increase of forward voltage, common energy bands in both regions are getting closer and at some voltage, no available states will exist at the opposite side and therefore no BTBT effect will occur. A further increase of the voltage bias will lead to thermionic carrier injection where electrons can pass over the tunneling potential barrier as shown in Figure 2.1(e). In this case, diffusion current and excess current start to dominate the total current. As shown in Figure 2.1(a), when a negative bias (reverse) is applied to the tunnel diode, electrons can tunnel from the valence band of the p-region to the conduction band of the n-region and the NDR effect is not observed. In the following sections, this reverse BTBT effect will be explained in more detail as it is presented as the main carrier injection mechanism of gated Tunnel diodes [Tunnel Field-Effect Transistors (TFETs)].

## 2.2 Band-to-Band Tunneling (BTBT) Current

In classical mechanics, carriers are confined by the potential walls between regions, and only those with excess energy higher than the barriers can escape from one region to another by thermionic emission. In contrast and as shown in Figure 2.2, in quantum mechanics a carrier can be represented by its wave function $\psi$ that does not terminate abruptly on a wall of finite potential height $U_0$ and therefore there is a non-zero probability of tunneling from one region to another through the barrier [3].

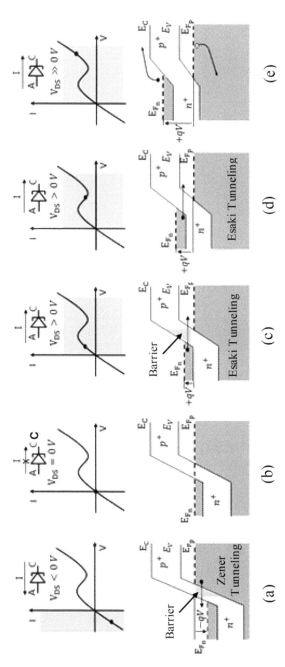

**Figure 2.1** Description of the tunneling effect in a diode with heavily doped *p-n* junctions. (a) Band-to-Band Tunneling (BTBT) current resultant from reverse biased diode; (b) thermal equilibrium; (c) BTBT current resultant from forward biased diode; (d) Decrease of BTBT current; (e) Diffusion and excess current with no BTBT effect.
*Source:* Adapted from [3].

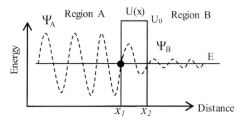

**Figure 2.2**    Wave function showing carrier tunneling through the barrier.
*Source:* Adapted from [3].

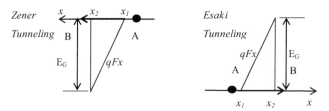

**Figure 2.3**    The triangular potential barrier seen by the tunneling carrier.

As explained in the previous section and as shown in Figure 2.3, in a tunneling device, two types of tunneling current can be identified: the Zener tunneling current (electrons tunneling from the valence band to the conduction band) and the Esaki tunneling current (electrons tunneling from the conduction band to the valence band).

In a tunneling device, the current is dependent on the tunneling transmission probability $T_{BTBT}$. The tunneling current is calculated using the Wentzel–Kramers–Brillouin (WKB) approximation that gives the transmission tunneling coefficient expressed in Equation (2.1) [3]:

$$I_{BTBT} \simeq T_{BTBT} = \frac{|\Psi_B|^2}{|\Psi_A|^2} \approx exp\left[-2\int_{x1}^{x2}|k(x)|dx\right] \qquad (2.1)$$

The distance-dependent term *k(x)* is the quantum wave vector of the carrier inside the barrier, and can be calculated as follows:

$$k(x) = \sqrt{\frac{2m^*}{\hbar}(E - U(x))} \qquad (2.2)$$

In Equation (2.2), $m^*$ is the effective mass of the carrier, $\hbar$ is the reduced Planck constant, $U(x)$ is the potential barrier energy and $E$ is the energy of the carrier. $U(x)$ is related with $E$, maximum electric field $F$ (assumed constant) at the tunneling junctions, and energy bang gap of the semiconductor $E_G$ as fllows:

$$U(x) = E + qFx \ (x_1 < x < x_2), \ (x_2 - x_1) = E_G/qF \quad (2.3)$$

Since $U(x) > E$, the wave vector $k(x)$ results in an imaginary number. Replacing Equation (2.3) in Equation (2.1) and considering the potential barrier shape shown in Figure 2.3, the tunneling transmission probability is given as:

$$T_{BTBT} \simeq exp\left[ -\frac{4\sqrt{2m^*}E_g^{3/2}}{3\hbar qF} \right] \quad (2.4)$$

Knowing the tunneling transmission factor $T_{BTBT}$, the tunneling current is then obtained by integrating the product of the number of available carriers in the originating region A and the probability of tunneling through the number of empty states in the destination region B over the range of overlapping energy states. The Zener $I_Z$ and Esaki $I_E$ tunneling current are calculated as follows:

$$I_Z(E) = I_{v \to c}(E)$$
$$= C1 \int_{E_v}^{E_c} F_v(E) \cdot N_v(E) \cdot [1 - F_c(E)] \cdot N_c(E) \cdot T_{v \to c}(E)$$
$$(2.5)$$

$$I_E(E) = I_{c \to v}(E)$$
$$= C1 \int_{E_c}^{E_v} F_c(E) \cdot N_c(E) \cdot [1 - F_v(E)] \cdot N_v(E) \cdot T_{c \to v}(E) \quad (2.6)$$

The term $C1$ is presented as a constant and $F_c$, $F_v$, $N_c$ and $N_v$ represent respectively the Fermi–Dirac distributions and densities of states in the respective corresponding regions. Kane derived in 1961 a solution for the tunneling current expression shown in Equations (2.5) and (2.6) [4]. The resultant BTBT generation rate $G_{BTB}$(current per unit area) can be presented as:

$$G_{BTBT} = A\frac{F^n}{E_G^{1/2}}exp\left( - B\frac{E_g^{3/2}}{F} \right) \quad (2.7)$$

In Equation (2.7) the two variables $A$ and $B$ are dependent on the device structure and material, and are usually used as fitting parameters during simulations in order to fit the experimental tunneling current-voltage characteristics. According to direct (no assistance from a phonon) or indirect tunneling, the exponential factor $n$ takes respectively a value of 2 and 2.5. The equation suggests that in order to achieve a large tunneling generation rate, a high electric field at the junctions and the use of small bandgap and mass materials are required. As will be shown in the following sections, the use of Ge and III–V materials in tunneling devices increases the tunneling current at similar bias when compared to silicon (Si)-based tunneling devices.

## 2.3 From Tunnel Diode to Gated p-i-n Structure

In this section, the most relevant evolutionary advances of tunneling devices are presented. The first works related to gated tunnel devices are mentioned, as also the most relevant changes performed in the device structure during the past decade. The use of low energy bandgap $E_G$ and mass materials $m^*$ in TFETs was shown to improve the electrical characteristics at low-voltage operation (sub-0.5 V). Simulated and experimental works performed by several groups are presented, as also the methodologies to increase the overall transistor performance.

### 2.3.1 First Observations of Tunneling in Gated Structures

Two decades after the demonstration of a tunneling device by Esaki, Quinn group at Brown University in 1978 was the first to propose a gated *p-n* structure over a $p^-$ substrate which they called a Surface Channel Tunnel Junction (SCTJ) transistor [5]. They proposed to replace the degenerated n-type source by a highly degenerated p-type source in the n-type metal-oxide-semiconductor field-effect transistor (MOSFET) (NMOS) device, keeping the p-type substrate as shown in Figure 2.4. They theorized that applying a large enough gate-to-source voltage would equally create a surface-inversion layer such as in the MOSFET; however, the conducting channel would be separated from the p-type source by a surface tunnel junction. A higher Fermi level in the source side than in the surface channel would allow electrons to tunnel from the source into unoccupied states of the drain region.

Later, in 1987 at Texas Instruments, Banerjee et al. observed the principle of Zener tunneling (electrons tunneling from valence to conduction band) in a planar Si-MOS device with three terminals, and a *p*-substrate [6]. Several characteristics were identified:

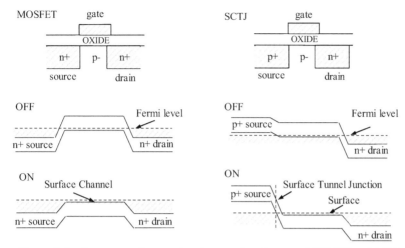

**Figure 2.4** Structure and energy band comparison between MOSFET and SCTJ.

- Saturated current in the output characteristics of the tunneling device. This behavior is related to the absence of the barrier-lowering effects presented in MOSFETs, allowing tunneling devices to be scalable;
- Non-linear trans-conductance and high output impedance related to the previous point. Such characteristics were expected to benefit non-linear analog circuit applications such as mixers or voltage reference sources;

One year later in 1988 at Hitachi Ltd., Takeda et al. proposed a tunneling device structure similar to that presented by Banerjee et al. called "Band-to-Band tunneling MOS device" ($B^2T$-$MOSFET$) [7]. They demonstrated that such a device shows no short-channel effect such as $V_{TH}$–lowering and little conventional hot carrier effects resulting in high scalability down to 0.1 μm. Other important characteristics were identified:

- Like conventional MOSFETs, at high temperatures, the tunneling leakage current is dominated by thermal generation-recombination;
- At large voltage $|V_{DS} > 5$ V| the tunneling current shows a small dependence on the temperature;

During the same year, the first vertical tunnel device was anticipated by Leburton et al. at the University of Illinois, United States [8]. They proposed a new tunneling device aimed at the possibility of operating as an ultrafast NDR three-terminal device. With a different carrier injection mechanism, they predicted a bipolar behavior of the transistor, where both holes and electrons were involved in the conduction mechanism. They called such new device "BiFET" (BiPolar TFET), with two novel features: the possibility to externally control with a third terminal the carrier injection in the tunnel junction, and the onset of the NDR effect.

In 1992, Baba at NEC Corporation independently proposed changes to the lateral tunneling device of Banerjee et al. using a gated *p-i-n* structure [9]. This new transistor was designed to use the gate to control the source-to-drain tunneling current, which he called Surface Tunnel Transistor (STT). Compared to the Banerjee's device, the gate of the proposed STT is not overlapped with the $p^+$ region, and no inversion layer is created. The STT was fabricated using GaAs by molecular-beam epitaxy (MBE), mesa etching, and regrowth techniques. At a low temperature of 77 K, the tunneling current was controlled by the gate, showing no saturation in the output characteristics.

Using the same GaSb-based STT structure in 1994, Uemura and Baba observed for the first time the NDR effect (Esaki tunneling) at room temperature [10]. Only three years later, the same effect was observed (at room temperature) in a Si-based tunneling device by Koga and Toriumi at Toshiba Corp. [11]. The *NDR* was observed at large gate-to-source voltage $V_{GS}$ (2.3–2.6 V) and low drain-to-source voltage $V_{DS}$ (0.25–0.35 V).

In 1995, Reddick and Amaratunga at Cambridge University were the first to demonstrate the Zener tunneling effect in a Si-based STT at room temperature [12]. The tunneling current was shown to be controlled by the gate, with low saturation effects. The device showed a large current gain (ratio of drive and leakage current) at large $V_{GS}$ ($I_{ON}/I_{OFF} = 10^6$ for $|V_{GS}| = 5$ V) but a small one at low voltage ($I_{ON}/I_{OFF} = 10^2$ for $|V_{GS}| = 1$ V).

### 2.3.2 Structural Improvements for Boosted Performance

In order to increase the current gain of tunneling devices at low-voltage operation, changes in the conventional device structures were required and proposed. In 2000, Hansch et al. at the University of the German Federal Armed Forces in Munich, Germany was the first to fabricate a vertical MOS gated tunneling transistor in silicon by means of Molecular Beam Epitaxy (MBE) [13]. As shown in Figure 2.5, the top region of the transistor (acting as the drain electrode) was formed by the deposition of a highly-doped boron delta-layer that provides the necessary abrupt *p-n* junction for the tunneling effect. A gate oxide was grown vertically with a thickness dimension of 20 nm. The device explored the Esaki tunneling effect, showing promising results at low voltage operation: a current gain of $10^3$ was shown at low supply voltages ($V_{SD} < -0.2$ V) and leakage current in the nA level.

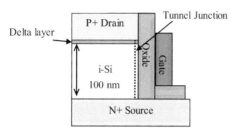

**Figure 2.5** Schematic view of the vertical Esaki-tunneling field-effect transistor (Esaki-FET).
*Source:* Adapted from [13].

In 2004, a work published by Bhuwalka at al. proposed changes in the Esaki-FET structure of Hansch. The boron delta layer was replaced by one with lower-energy bandgap (heavily doped 3 nm SiGe layer) in order to reduce the tunnel barrier width at the source-channel junction and therefore increase the drive current [14]. The structure was simulated for different channel lengths (100 nm, 70 nm, and 50 nm) and the author concluded that increasing the Ge mole fraction in the SiGe layer decreases the $V_{TH}$ in the current-voltage characteristics of the tunneling device but consequently increases the leakage current, thus decreasing the current gain. This characteristic was shown to be more intense at a lower channel dimension, which is presented as a problem for scaled devices. The vertical device with such topology was fabricated, showing promising results: Zener tunneling at room temperature was observed, with a perfect saturation in the output characteristics of the device [15]. Despite the low on-current (sub-$\mu$A for $V_{GS} = 8$ V and $V_{DS} = 1$ V), a current gain with 5 orders of magnitude demonstrated the potential of tunneling devices for low-power applications.

During the same year, Wang et al. at the Technical University of Munich fabricated and demonstrated for the first time, complementary silicon tunneling transistors on the same silicon substrate [16]. They showed that high doping concentration in both n-type and p-type surfaces ($> 2 \times 10^{20} \text{cm}^{-3}$) enhances the Zener tunneling current density and can improve the threshold voltage of tunneling devices, or as they called it, TFETs. For an n-type TFET (source is $p+$ and drain is $n+$) and a supply voltage of 3 V, a current gain of $10^5$ was achieved, although with a poor inverse sub-threshold slope (SS) of 471 mV/dec. This value was shown to be far from the ideal and limited SS of 60 mV/dec of thermionic devices such as conventional MOSFETs.

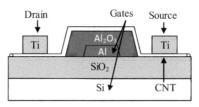

**Figure 2.6**    CNT-based TFET.

*Source:* Adapted from [17].

For a *p*-type TFET (source is *n+* and drain is *p+*) a current gain of about $10^6$ can be achieved with a $V_{GS}$ voltage of –3 V. The *SS* of the p-TFET was shown to be 106 mV/dec. Both devices were characterized by a low leakage current ($\sim$pA/$\mu$m).

Still in 2004, Appenzeller et al. at IBM (New York, United States) and ISG (Jülich, Germany) were the first to observe the Band-to-Band Zener tunneling effect in a dual-gated Carbon Nanotube (CNT) Field Effect Transistor (CNT-TFET) [17]. An SS value lower than 60 mV/dec at room temperature was for the first time demonstrated due to the controlled tunneling behavior. At a $V_{GS\,Al}$ = –1.5 V and $V_{DS}$ = –0.5 V, the *CNTTFET* presented a drive current of approximately 0.1 $\mu$A and a current gain of approximately $10^7$. In order to create the necessary energy bands, two independent gates (Si and Al) were used, both located underneath the CNT, as shown in Figure 2.6.

### 2.3.3 Tunnel FET Evolution over the Past Decades

The last decade has been fertile in works (both simulated and experimental) related to the improvement of TFET structures, in order to achieve similar drive currents to that of MOSFETs, maintaining a low leakage current and an SS below 60 mV/dec at room temperature.

In 2007, Boucart and Ionescu at EPFL studied by simulations the characteristics of a double-gate Tunnel FET (DG-TFET) such as the one shown in Figure 2.7, using high-k gate dielectrics with optimized silicon body thickness [18].

The transmission tunneling probability previously presented in Equation (2.4) can be adapted to different device structures as shown by Equation (2.8). The screening length $\triangle\Phi$ and natural length $\lambda$ are represented in Figure 2.8. In Table 2.1, the natural length expression for different gate configurations is shown, suggesting that the use of high-k dielectrics in multi-gate configurations can increase the tunneling current [19]. This is due to the improved electrical coupling between the gate and the tunneling junction caused by the increased gate capacitance.

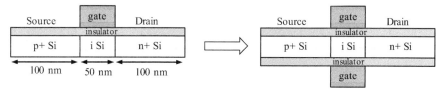

**Figure 2.7** Single-gate and double-gate structure TFET with high-k gate dielectric. *Source:* Adapted from [18].

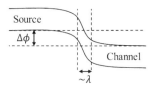

**Figure 2.8** Energy band cross section of the TFET.

$$T(E) \simeq exp\left(-\frac{4\lambda\sqrt{2m^*}E_g^{3/2}}{3\hbar q\left(\triangle\Phi + E_g\right)}\right)\triangle\Phi \qquad (2.8)$$

In Boucart's work, the performance of a DG-TFET designed with a $SiO_2$ gate dielectric ($\varepsilon_{ox} = 3.9$) was compared with a DG-TFETs designed with $HfO_2$ ($\varepsilon_{ox} = 21$) and $ZrO_2$ ($\varepsilon_{ox} = 29$) considering oxide thicknesses of 3 nm.

**Table 2.1** Natural length for different device configurations

| | |
|---|---|
| Single-gate | $\lambda = \sqrt{\dfrac{\varepsilon_{si}}{\varepsilon_{ox}}t_{si}t_{ox}}$ |
| Double-gate | $\lambda = \sqrt{\dfrac{\varepsilon_{si}}{2\varepsilon_{ox}}t_{si}t_{ox}}$ |
| Gate-all-around (Square Channel Cross Section) | $\lambda = \sqrt{\dfrac{\varepsilon_{si}}{4\varepsilon_{ox}}t_{si}t_{ox}}$ |
| Gate-all-around (Circular Channel Cross Section) | $\lambda = \sqrt{\dfrac{2\varepsilon_{si}t_{si}^2 ln\left(1+\dfrac{2t_{ox}}{t_{si}}\right) + \varepsilon_{ox}t_{si}^2}{16\varepsilon_{ox}}}$ |

*Source:* Adapted from [19].

With a double-gate (DG) structure and high-k dielectrics, the tunneling device based on the Zener effect was characterized by improved on-current and lower SS values compared to the single-gate (SG) TFET structure. On-current values of 0.23 mA at $V_{GS} = 1.8$ V and an SS value of 57 mV/dec were achieved for an optimum silicon body layer thickness of 7 nm and a gate length of 50 nm. The authors concluded that the current gain of $10^{11}$ and the leakage current lower than 1 fA/$\mu$m makes the TFET device a promising candidate to complement or replace the MOSFET technology, particularly in the Low Standby Power (LSTP) category.

Still in 2007, Choi et al. from Berkeley demonstrated in 2007 the first TFET device in a Silicon on Insulator (SOI) (70 nm n-channel) with an SS lower than 60 mV/dec at room temperature [20]. With a gate dielectric of 2 nm ($SiO_2$) and an SOI layer of 60 nm, an SS value of 52.8 mV/dec was demonstrated. At a power supply voltage of 1 V, the on-current and leakage current achieved were respectively 12.1 $\mu$A/$\mu$m and 5.4 nA/$\mu$m. In order to increase the current gain, the authors proposed the use of lower energy bandgap materials, dielectrics with lower equivalent oxide thickness (EOT) and a more abrupt source doping profile (values not specified).

Following this work, Mayer et al. at CEA-LETI reported for the first time experimental investigations on SOI, heterostructure $Si_{1-x}Ge_xOI$ (x = 15% or 30%) and GeOI-based Tunnel FETs [21] in a fully depleted SOI CMOS process flow using high-k metal-gate stack integration ($HfO_2$ and TiN). The authors showed that at room temperature, the SOI-TFET is characterized by a very low leakage current of $\sim$30 fA/$\mu$m (at $|V_{DS}| = 0.6$ V), an SS of 42 mV/dec over a wide $|V_{DS}|$ range (0.1–1 V) and a low on-current of $\sim$50 nA/$\mu$m at $|V_{DS}| = 1$ V and $|V_{GS}| = 2$ V. The on-current was shown independent of gate length (BTBT injection is not limited by carrier transport as conventional thermal devices as MOSFETs). Compared to the use of SOI, GeOI-based TFETs showed increased on-current ($\times$2700 for p-type and $\times$335 for n-type) due to the lower energy bandgap of Ge ($E_G = 0.66$ eV) compared to Silicon ($E_G = 1.1$ eV). Despite the improved on-current, the leakage current of GeOI-TFET increased by 5 orders of magnitude when compared to SOI-TFET. A heterojunction SiGeOI-TFET was shown to improve the on-current by approximately 1 order of magnitude, with a consequent increase of leakage current on both n- and p-type SiGeOI-TFETs.

In 2009, Luisier and Klimeck published an interesting paper studying an InAs-TFET with SG and DG ultra-thin body (UTB), both with a gate length of 20 nm, and a gate-all-around nanowire (GAA NW) [22]. They concluded that a reduced SS can be achieved if the electrostatic potential under the gate contact is very well controlled, finding that GAA-NWs can keep an SS lower than 60 mV/dec with diameters larger than 10 nm, while the bodies in DG and SG must be scaled down to 7 nm and 4 nm respectively.

During the same year, Moorkejea et al. at Pennsylvania State University fabricated an heterostructure TFET, using groups III–V materials in the source region (InGaAs) and a high-k gate dielectric (Al$_2$O3), expecting to improve the on-current due to the smaller energy bandgap and electron mass of such materials compared to Ge and Si. For a vertical gate length of 100 nm, a current gain of $10^4$ was achieved with an on-current of 20 $\mu$A/$\mu$m at $V_{DS} = 0.75$ V and $V_{GS} = 2.5$ V [23].

In 2011, Dewey and his group at Intel fabricated a vertical tunneling device based on *InGaAs* (60 nm $p+$ In$_{0.53}$Ga$_{0.47}$As source, 100 nm intrinsic In$_{0.53}$Ga$_{0.47}$As channel and thick $n+$ In$_{0.53}$Ga$_{0.47}$As drain) [24]. In order to reduce the parasitic leakages, the gate and source pads were isolated via mesa etch with metal air bridges as shown in Figure 2.9. A 6-nm In$_{0.7}$G0$_{.3}$aAs "pocket" was grown between the source and channel regions, thus allowing for a reduction in the tunneling barrier height at the source-channel interface. Such configuration allowed for an increased on-current and the lowest SS value reported at that time by III–V TFETs ($\sim$60 mV/dec). The highest on-current ($\sim$7$\mu$A/$\mu$m at $V_{GS} = 0.8$ V and $V_{DS} = 0.3$ V) was achieved considering a gate dielectric with *EOT* = 1.1 nm and a source doping concentration of $1\times10^{20}$/cm$^{-3}$. The current gain was shown to be approximately $10^5$.

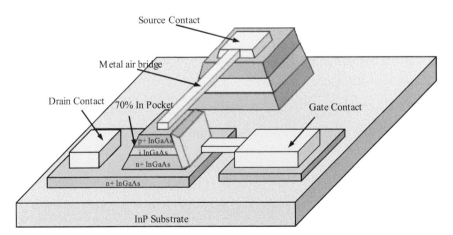

**Figure 2.9** Schematic of InGaAs-TFET.

*Source:* Adapted from [24].

During the same year, the steepest SS value in Si-based TFETs was demonstrated by Gandhi et al. at A*Star, Singapore [25]. A value of 30 mV/dec was achieved with a CMOS-compatible vertical GAA structure. The leakage current achieved was in the order of fA/$\mu$m. However, the on-current presented low values (5 decades of magnitude larger than the off-current considering a power supply voltage of 1.2 V).

The tunneling effect in a multi-gate TFET configuration was also demonstrated. In order to improve the transistor characteristics of the TFET, in 2011 Leonelli et al. at IMEC, Leuven, Belgium, fabricated a multi-gate TFET device in SOI. The configuration was characterized by two gate dielectrics (high-k $HfO_2$ and low-k $SiO_2$), enabling large on-current values compared to other structures (46 $\mu$A/$\mu$m at $|V_{DD}|$ = 1.2V), SS of 100 mV/dec, and a current gain of $10^6$ [26].

The last few years have been fertile in works demonstrating the increased performance of tunneling devices using materials from groups III and V of the periodic table due to their low mass and energy bandgap. In 2012, Zhou et al. at the University of Notre Dame, United States presented their vertical heterojunction TFET based in III–V materials (GaSb–InAs) and two high-k gate dielectric ($Al_2O_3$ – $HfO_2$) in which the gate field was aligned with the tunneling direction, resulting in a record on-current at that time (180 $\mu$A/$\mu$m) at low voltage values ($V_{GS}$ = $V_{DS}$ = 0.5 V) [27]. Despite the large drive-current the current gain was shown to be low, about $10^4$ (large leakage current) and the SS value was presented as 200 mV/dec (considering a high-k gate dielectric ($Al_2O_3$/$HfO_2$, EOT = 1.3 nm).

Also in 2012, Tomioka et al. at Hokkaido University, Japan reported a III–V NW/Si heterojunction TFET ($n$+ InAs drain and intrinsic channel, $p$+ Si source) with a surrounding gate architecture and high-k dielectrics [28]. The lowest SS so far (21 mV/dec) was demonstrated for a $V_{DS}$ range of 0.1–1 V and a NW diameter of 30 nm. The on-current was shown to be approximately 1 $\mu$A/$\mu$m at $V_{DS}$ = 1 V (current gain is $10^6$).

In 2013, Noguchi et al. at the University of Tokyo, Japan demonstrated their planar InGaAs TFET, showing a record SS of 64 mV/dec in III–V planar devices (with EOT of 1.4 nm), with a current gain of more than $10^6$ ($I_{ON}$ = 5 $\mu$A/$\mu$m at $V_{DS}$ = 0.15 V) [29]. The authors concluded that a formation of a defect-free source junction with steep impurity profiles is mandatory to fully develop the potential of planar type TFETs.

In 2015, Pandey et al. at the Penn State University, United States demonstrated a complementary "all III–V" heterojunction vertical Tunnel FET (*HVTFET*) with record performance at $|V_{DS}|$ = 0.5 V [30]. The n-type *TFET* showed an on-current of 275 μA/μm and a current gain of $3 \times 10^5$ (SS = 55 mV/dec) while the *p*-type TFET presented an on-current of 30 μA/μm with a current gain of $10^5$ (SS = 115 mV/dec). All these results were measured at room temperature. For the GaAs$_{0.35}$Sb$_{0.65}$ p-TFET channel, a 3.5-nm HfO$_2$ gate dielectric was considered while a 4-nm ZrO$_2$ high-K dielectric was considered for the InGaAs n-TFET channel.

Regarding the state of the art presented in this section, Figure 2.10 summarizes the most important evolutionary steps of gated tunneling devices.

### 2.3.4 Directions for Further Improvements in Tunneling Devices

According to the state of the art and as summarized in Figure 2.11, tunneling devices fabricated with groups III–V materials are able to provide larger on-current or drive-current values when compared to Si-based counterparts. This is explained due to the lower energy band gap E$_G$ and mass of III–V components, thus allowing for an increased tunneling transmission probability according to (2.4). Despite larger on-current magnitudes, III–V-based tunneling devices still present a large leakage current that not only degrades the SS of the device but also limits the current gain.

Some publications suggest that Trap-Assisted Tunneling (TAT), Band Tails due to heavy-doping, interface roughness, and interface traps at the high-k dielectric/semiconductor interface are the main non-ideality factors that contribute to large SS values [31–34].

- Trap-assisted Tunneling: TAT is presented as an additional thermal electron-hole pair-generation mechanism in TFETs that sets prior to BTBT due to lower tunnel barrier. Trap levels in the device are due to imperfections in the crystal periodicity such as lattice defects or impurity atoms. Bulk traps, traps at the material interface, and traps at the oxide contribute to the TAT current by modifying the electrostatics inside the device. In heterojunction TFETs, it was shown that traps located near the source end of the channel with levels closer to the conduction band are more likely to degrade the SS. This is because a charge trap near the source-channel interface alters the junction electric field, thus affecting the tunneling rate [31];
- Band tails: The density of states (DOS) extending in to the bandgap due to high doping density is another factor that degrades the performance of *TFETs*. In tunneling devices, high electric fields are required in the source-channel region in order to increase the tunneling

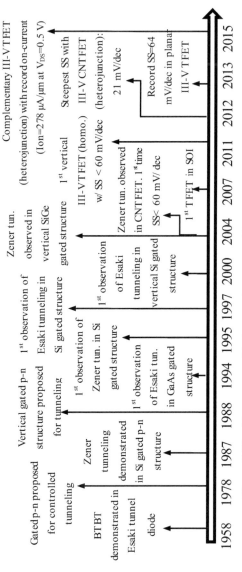

**Figure 2.10**   Chronogram with the most important achievements of tunneling devices.

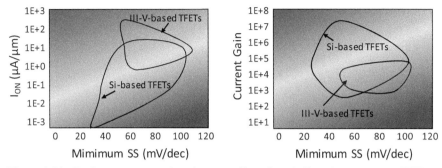

**Figure 2.11** Performance comparison between silicon-based (Si) and groups III–V TFETs.

transmission probability [see Equation (2.4)]. To support high electric fields, high source doping is required, giving rise to band tails that decay exponentially into the bandgap. The consequent non-homogeneous distribution of dopant atoms in the tunneling regions creates different local potentials, affecting the electric field distribution over the channel;

- Interface roughness: A rough oxide–semiconductor interface cause random fluctuations of the boundary wall of the triangular potential shown in Figure 2.3. This results in extra-tunneling paths for carriers to tunnel from one region to the other.

In order to counteract such defects, further investigation in III–V and novel materials that currently present high bulk and interface defects is required in order to achieve the maturity level of silicon and steep inverse SSs. The use of novel materials in TFETs such as Graphene or Molybdenum Disulfide ($MoS_2$) has gained momentum recently and is presented as an option to improve the tunneling device performance. In [35], the authors have shown by simulations that a TFET based on a 2D graphene bilayer can achieve a steep SS as low as 20 $\mu$V/dec and a large on-current of 0.8 mA/$\mu$m. The large current is explained by the narrow extrinsic bandgap of the material ($\sim$0.3 eV) while the steep SS is explained by an all-electrical doping process (instead of chemical) in the source and drain contact, which suppresses the band tailing and TAT mentioned in this section. The current gain is shown to be $10^5$ for a power-supply voltage of 150 mV.

In [36], the authors have demonstrated by simulations a steep SS of 3.9 mV/dec in a TFET designed with a highly doped germanium source and atomically thin $MoS_2$ as the channel, in a vertical structure. Despite the low current (sub-10 $\mu$A), the tunneling device can operate with a low

$V_{DS}$ of 0.1 V. A current gain of more than $10^8$ was observed by the authors concluding that such a device could be applied in future low-power integrated circuits.

### 2.3.5 A Brief Discussion of the Tunneling Device State of the Art

According to the review performed in this chapter, the results presented by several groups point to some conclusions:

- The BTBT carrier-injection mechanism based on the Zener tunneling effect will enable switches with an inverse SS below the limited 60 mV/dec at room temperature of conventional thermal dependent MOSFETs;
- With the eventual fabrication maturity of novel materials in groups III–V (e.g. InGaAs, GaSb, InAs), Graphene and MoS$_2$, and further improvements in the development of defect-free CNT, one can expect tunneling devices with a very low leakage current (sub-0.1 pA/$\mu$m);
- Tunneling devices can operate at ultra-low voltage (sub-0.25 V);
- Due to the BTBT mechanism, TFETs are scalable as the carrier injection is strongly dependent on the tunneling barrier between the source and channel regions;
- The on-current is still low compared to conventional MOSFETs that presents drive currents on the order of mA/$\mu$m;

With the mentioned points, the tunneling device is presented as a promising candidate for ultra-low-power applications. The large current gain at low voltage operation will enable the design of more energy-efficient circuits compared to those existing today, and therefore, the exploration of such devices in circuits for energy-harvesting applications is a natural choice, requiring further investigation.

### References

[1] Esaki, L. "Long journey into tunneling," in *Proc. IEEE*, vol. 62, no. 6, pp. 825–831, 1974.
[2] General Electric Research laboratory, "Tunnel Diodes." Available at: http://n4trb.com/AmateurRadio/SemiconductorHistory/GE_Tunnel_Diodes.pdf, 1959.

[3] Sze, S. M., and Ng, K. K. "Physics of Semiconductor Devices," 3rd Edn. Hoboken, NJ: John Wiley & Sons, 2007.

[4] Kane, E. O. "Theory of tunneling," *Appl Phys. Lett.* 32, 83–91, 1961.

[5] Quinn, J., Kawamoto, G., and McCombe, B. "Subband spectroscopy by surface channel tunneling," *Surface Science* 73, 190–196, 1978.

[6] Banerjee, S. et al. "A new three-terminal tunnel device," *IEEE Electron Device Letters*, vol. 8, pp. 347–349, 1987.

[7] Takeda, E., Matsuoka, H., Igura, Y., and Asai, S. "A band to band tunneling MOS device ($B^2$T-MOSFET)," in *IEDM Tech. Dig.*, pp. 402–405, 1988.

[8] Leburton, J. P., Kolodzey, J., and Biggs, S. "Bipolar tunneling field-effect transistor: A three-terminal negative differential resistance device for high-speed applications," *Appl. Phys. Lett.* 52, 1608–1620, 1988.

[9] Baba, T. "Proposal for surface tunnel transistors," *Jpn. J. Appl. Phys.* 31, 455–457, 1992.

[10] Uemura, T., and Baba, T. "First observation of negative differential resistance in surface tunnel transistors", *Jpn. J. Appl. Phys.* 33, 207–210, 1994.

[11] Koga, J., and Toriumi, A. "Negative differential conductance at room temperature in three-terminal silicon surface junction tunnel transistor", *Appl. Phys. Lett.* 70, 16, 2138–2140, 1997.

[12] Reddick, W., and Amaratunga, G. "Silicon surface tunnel transistor,"*Appl. Phys. Lett.* 67, 494–496, 1995.

[13] Hansch, W., Fink, C., Schulze, J., Eisele, I. "A vertical MOS-gated Esaki tunneling transistor in silicon," *Thin Solid Films* 369, 387–389, 2000.

[14] Bhuwalka, K. K., Schulze, J., Eisele, I. "Scaling issues of n-channel vertical tunnel FET with $\delta$ p+ SiGe layer," in *Device Research Conference IEEE*, vol. 1, pp. 215–16, 2004.

[15] Bhuwalka, K. K., Sedlmaier, S., Ludsteck, A. K., Tolksdorf, C., Schulze, J., and Eisele, I. "Vertical tunnel field-effect transistor," in *IEEE Trans. Electron Devices*, vol. 51, no. 2, pp. 279–282, 2004.

[16] Wang, P.-F., Hilsenbeck, K., Nirschl, T. H., Oswald, M., Stepper, C. H., Weis, M., et al., "Complementary tunneling transistor for low power application," *Solid-State Elec.*, vol. 48, pp. 2281–2286, 2004.

[17] Appenzeller, J., Lin, Y.-M., Knoch, J., and Avouris, P. H. "Band-to-Band tunneling in carbon nanotube field-effect transistors," *Phys. Rev. Lett.* 93:19, 2004.

[18] Boucart, K., and Ionescu, A. M. "Double-gate tunnel FET with high-k gate dielectric," in *IEEE Trans. Electron Devices*, vol. 54, no. 7, pp. 1725–1733, 2007.

[19] J.-P. Colinge, Ed., *"FinFETs and Other Multi-Gate Transistors"*, New-York, NY: Springer, 2008.

[20] Choi, W. Y., Park, B.-G., Lee, J. D., Liu, T.-J. K. "Tunneling field-effect transistors (TFETs) with subthreshold swing (SS) less than 60 mV/dec," in *IEEE Electron Device Letters*, vol. 28, no. 8, pp. 743–745, 2007.

[21] Mayer, F., Le Royer, C., Damlencourt, J.-F., Romanjek, K., Andrieu, F., Tabone, C., et al. "Impact of SOI, Si1-xGexOI and GeOI substrates on CMOS compatible Tunnel FET performance," in *IEDM Tech. Dig.*, pp. 163–166, 2008.

[22] Luisier, M., and Klimeck, G. "Atomistic Full-Band Design Study of InAs Band-to-Band Tunneling Field-Effect Transistors," in *IEEE Elec. Dev. Letters*, vol. 30, no. 6, pp. 602–604, 2009.

[23] Mookerjea, S., Mohata, D., Krishnan, R., Singh, J., Vallett, A., Ali, A., et al. "Experimental demonstration of 100nm channel length $In_{0.53}Ga_{0.47}As$-based vertical inter-band tunnel field effect transistors (TFETs) for ultra low-power logic and SRAM applications," *Int. Electron Devices Meeting*, pp. 1–3, 2009.

[24] Dewey, G., Chu-Kung, B., Boardman, J., Fastenau, J. M., Kavalieros, J., Kotlyar, R., et al. "Fabrication, characterization, and physics of III–V heterojunction tunneling field effect transistors (H-TFET) for steep sub-Threshold swing," in *Int. Electron Devices Meeting*, pp. 33.6.1–33.6.4, 2011

[25] Gandhi, R., Chen, Z., Singh, N., Banerjee, K., and Lee, S. "Vertical Si-nanowire n-type tunneling FETs with low subthreshold swing (= 50 mV/decade) at room temperature," in *IEEE Electron Device Letters*, vol. 32, no.4, pp. 437–439, 2011.

[26] Leonelli, D., Vandooren, A., Rooyackers, R., De Gendt, S., Heyns, M. M., and Groeseneken, G. "Drive current enhancement in p-tunnel FETs by optimization of the process conditions," *Solid-State Elec.* 65–66, 28–32, 2011.

[27] Zhou, G., Lu, Y., Li, R., Zhang, Q., Liu, Q., Vasen, T., et al. "InGaAs/InP tunnel FETs with a subthreshold swing of 93 mV/dec and $I_{ON}/I_{OFF}$ ratio near $10^6$," in *IEEE Electron Device Letters*, vol. 33, no. 6, pp. 782–784, 2012.

[28] Tomioka, K., Yoshimura, M., and Fukui, T. "Steep-slope tunnel field-effect transistors using III–V nanowire/Si heterojunction," in *Symposium on VLSI Technology (VLSIT)*, pp. 47–48, 2012.

[29] Noguchi, M., Kim, S., Yokoyama, M., Ji, S., Ichikawa, O., Osada, T., et al. "High Ion/Ioff and low subthreshold slope planar-type InGaAs tunnel FETs with Zn-diffused source junctions," in *2013 IEEE International Electron Devices Meeting*, pp. 28.1.1–28.1.4, 2013.

[30] Pandey, R., Madan, H., Liu, H., Chobpattana, V., Barth, M., Rajamohanan, B., et al.. "Demonstration of p-type In0.7Ga0.3As/GaAs0.35Sb0. 65 and n-type GaAs0.4Sb0.6/In0.65Ga0.35As complimentary heterojunction vertical tunnel FETs for ultra-low power logic," in *Symposium on VLSI Technology (VLSI Technology)*, pp. T206–T207, 2015.

[31] Pandey, R., Rajamohanan, B., Liu, H., Narayanan, V., and Datta, S. "Electrical noise in heterojunction interband tunnel FETs," in *IEEE Trans. on Electron Devices*, vol. 61, no. 2, pp. 552–560, 2014.

[32] Schenk, A., Sant, S., Moselund, K., and Riel, H. "III–V-based hetero tunnel FETs: A simulation study with focus on non-ideality effects," in *Joint International EUROSOI Workshop and International Conference on Ultimate Integration on Silicon*, pp. 9–12, 2016.

[33] Avci, U. E., Morris, D. H., and Young, I. A. "Tunnel field-effect transistors: Prospects and challenges," in *IEEE Journal of the Elec. Devices Society*, vol. 3, no. 3, pp. 88–95, 2015.

[34] Lu, H., and Seabaugh, A. "Tunnel field-effect transistors: State-of-the-Art," in *IEEE Journal of the Electron Devices Society*, vol. 2, no. 4, pp. 44–49, 2014.

[35] Alymov, G., Vyurkov, V., Rhyzii, V., and Svintsov, D. "Abrupt current switching in graphene bilayer tunnel transistors enabled by Van Hove singularities," *Scient. Rep.* 6:24654, 2016.

[36] Sarkar, D., Xie, X., Liu, W., Cao, W., Kang, J., Gong, Y., et al. "A sub-thermionic tunnel field-effect transistor with an atomically thin channel," *Nature* 526, 91, 2015.

# 3

# Tunnel FET: Physical Properties

In this chapter, a study on the Tunnel Field-Effect Transistors (TFET) current-voltage dependence on several physical parameters is performed considering devices designed with different materials (Si, Ge, and InAs). As TFETs present particular electrical characteristics, this chapter aims to identify key parameters in order to improve the TFET performance in circuit applications that require low and large intrinsic resistance at forward and reverse bias conditions respectively.

## 3.1 Thermionic Injection vs. BTBT

Unlike thermionic devices such as metal-oxide-semiconductor field-effect transistors (MOSFETs), the TFET is designed as a reverse-biased gated $p$-$i$-$n$ diode. As shown in Figure 3.1, the source of an n-type TFET presents a different doping type than that of an n-type MOSFET (NMOS) (the $n^+$ source of MOSFETs is replaced by a $p^+$ source). In p-type TFETs and p-type MOSFETs (PMOS), the source presents different doping types: $n^+$ for TFETs and $p^+$ for MOSFETs. With different doping structures, TFETs and MOSFETs present different energy-band diagrams and different carrier-injection mechanisms.

As shown in Figure 3.2, the switching mechanism of conventional MOSFETs is based on the injection of carriers from a thermally broadened Fermi distribution in the source region, over a potential barrier in the conduction band of the channel region. For an NMOS, a positive gate voltage (in relation to the source) in the channel region moves down the energy bands and electrons move from the source region to the drain region. As the Fermi distribution broadening in the source region is temperature dependent, the increase of drain current $I_{DS}$ with the applied $V_{GS}$ is also temperature dependent. The inverse sub-threshold slope (SS) of current in MOSFETs

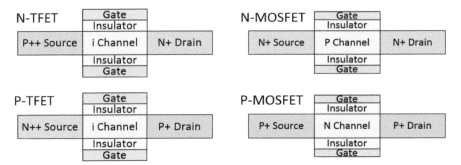

**Figure 3.1**    Double-gate TFET and double-gate MOSFET structure.

**n-MOSFET, Thermionic injection**

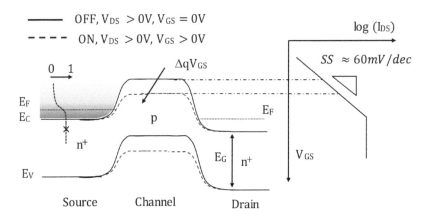

**Figure 3.2**    Energy-band diagram of n-MOSFET and resultant $I_{DS}$–$V_{GS}$ characteristics.

can be approximated by Equation (3.1), where $C_{OX}$, $C_{DEPL}$, and $C_{INT}$ are respectively the oxide, depletion, and interface capacitances. If the gate presents a good electrostatic control over the channel, i.e. $C_{OX} \gg C_{DEPL} + C_{INT}$, then the value of SS tends to its limiting value of 60 mV/dec at room temperature [1].

$$SS = \left[\frac{\partial \log \rightarrow (I_{DS})}{\partial V_{GS}}\right]^{-1} = \frac{\ln(10)\,k_B T}{e}\left(1 + \frac{C_{DEPL} + C_{INT}}{C_{OX}}\right) \quad (3.1)$$

$$SS = \frac{\ln(10)\,kBT}{e} \approx 60mV/dec \quad (3.2)$$

Due to the limited SS of MOSFETs, an increase of current by one order of magnitude requires at least a voltage of 60 mV in the gate. As an example, if one requires a thermionic device with a current gain ($I_{ON}/I_{OFF}$) of $10^5$, a supply voltage of at least 300 mV is required (at room temperature). Therefore, in order to enable efficient low-power circuits with reduced power consumption (by reducing the power supply voltage) and heat dissipation, a steeper slope in the $I_{DS}$–$V_{GS}$ characteristics of the device is required.

In Figure 3.3, the carrier injection mechanism of n-type and p-type TFETs is presented. The carrier transport of TFETs is based on Band-to-band-Tunneling (BTBT) instead of thermal emission over a potential barrier as previously shown by MOSFETs [1–2].

In *n*-TFETs the $p^+$ type source is heavily doped and the Fermi level is located below the energy of the valence band. When the device is operating in the off-state, i.e. $V_{GS} = 0$ V, no tunneling takes place due to the large potential barrier seen by electrons at the source side. By applying a positive gate-to-source voltage ($V_{GS}$), the valence and conduction band in the channel region are moved down and an energy overlap is created (energy window or screening length $\Delta\Phi$). This energy window enables electrons from the source side to tunnel through the channel to the drain side by the Zener effect. A similar phenomenon occurs in p-type TFETs. By applying a negative $V_{GS}$, the energy bands in the channel are moved up and an energy window is created at the source–channel interface, allowing holes to tunnel from the conduction band of the source region to the valence band of the drain region.

As explained in Chapter 2, the BTBT current is proportional to the tunneling transmission probability of carriers to tunnel from one region to the other. As expressed in Equation (3.3), larger $\Delta\Phi$ resultant from larger $|V_{GS}|$ magnitudes result in larger transmission probabilities and therefore larger BTBT current [3]. In TFETs, the current saturates at large $V_{GS}$ values due to the saturation of the source injection.

$$I_{BTBT} \propto T(E) \approx exp\left(-\frac{4\lambda\sqrt{2m^*}E_g^{3/2}}{3\hbar q\left(\triangle\Phi + E_g\right)}\right)\triangle\Phi \qquad (3.3)$$

As opposed to MOSFETs, the confined energy window of TFETs acts as a filter for carriers in the source region, cutting off the high and low energy tail of the Fermi distribution function. Consequently, the BTBT carrier injection mechanism of TFETs becomes independent of the broadening of the Fermi function with temperature. Therefore, the SS of TFETs is highly dependent on the energy window $\Delta\Phi$, changing according to the $V_{GS}$ magnitude.

n-TFET, electron BTBT

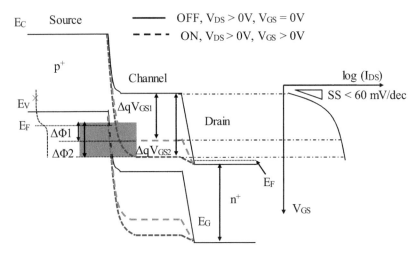

p-TFET, hole BTBT

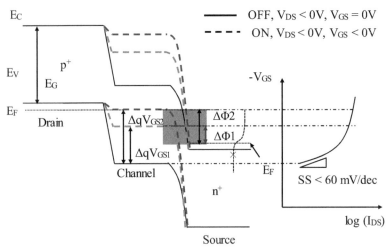

**Figure 3.3**    Band diagram of electron and hole BTBT in respective *n* and p-type TFETs and resulting $I_{DS}$–$V_{GS}$ characteristics.

This dependence has important implications on the TFET performance since the SS is not constant during the $V_{GS}$ range. This characteristic is further explained in Section 3.2.2. As expressed in Equation (3.4), SS in TFETs does not present a first-order temperature dependence such as MOSFETs, and therefore an SS value below 60 mV/dec at room temperature is theoretically possible [4].

$$SS \approx \frac{ln\,(10)}{e}\Delta\Phi \tag{3.4}$$

In TFETs, the SS is degraded due to the increase of leakage current consequent of several leakage carrier injection mechanisms [2]. Gate leakage through the high-k gate stack and thermionic emission over the built-in potential are mechanisms not only observed in conventional MOSFETs but also in TFETs. As shown in Figure 3.4(a), heavily doped source and drain regions can result in Shockley–Read–Hall generation. As illustrated in Figure 3.4(b), TFETs designed with short channels (sub-20 nm) can allow direct tunneling and defect-assisted tunneling, even without any bias applied to the gate. As shown in Figure 3.4(c), if the drain-to-source bias is larger than $E_G/q$, hole BTBT can occur at the drain-channel interface. Lowering the drain doping profile in order to extend the drain-channel tunnel junction, or design TFETs with different materials (heterojunction with $E_G$ in the source larger than $E_G$ in the drain and channel) are presented as two solutions to avoid the ambipolar characteristic shown in Figure 3.4(c).

## 3.2 Impact of Physical Properties in the TFET Performance

In this section, the results of a simulated TFET device designed with Atlas device simulator from Silvaco [5] are presented. The impact of several physical parameters on the electrical performance of the TFET, such as doping

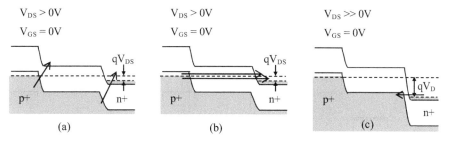

**Figure 3.4** Energy-band diagrams of n-type TFETs showing leakage mechanism during the off-state. (a) Shockley–Read–Hall generation in the source ($p^+$) and drain ($n^+$) regions; (b) direct and defect-assisted tunneling; (c) hole injection at the drain-channel interface.

*Source*: Adapted from [2].

levels in the drain and source regions, gate dielectrics with different permittivity and thickness dimensions, body thickness, and body materials is analyzed. At this point, it is important to mention that the right qualitative trend results are captured by the device simulator; however the quantitative predictions still requires further validation and calibration with experimental data.

### 3.2.1 Device Structure and Applied Model

In Figure 3.5, the structure of the simulated TFET device is shown. In order to improve the gate control over the channel, a double-gate configuration is considered. For simulation purposes the TFET is designed as an n-type configuration, i.e., with $p^+$ doping type in the source and $n^+$ doping type in the drain. Both source and drain regions are simulated with a length of 100 nm. The channel/gate length is simulated as 20 nm.

A non-local BTBT model was chosen prior to a local BTBT. As shown in Figure 3.6, the non-local model does not depend on the electrical field at individual mesh points, but rather on energy band diagrams calculated along cross-sections through the device. In contrast, the local BTBT model uses equations that assume a constant electric field over the tunneling length.

The tunneling current density at a given perpendicular energy $E$ for all values of $E$ between the energy range shown in Figure 3.6 and for each tunneling slice in the tunneling regions is calculated by Equation (3.5) [5]. In the expression, $E_{Fr}$ and $E_{Fl}$ are the quasi-Fermi levels belonging to the majority carrier at the relevant side of the junction. As an example, in Figure 3.6

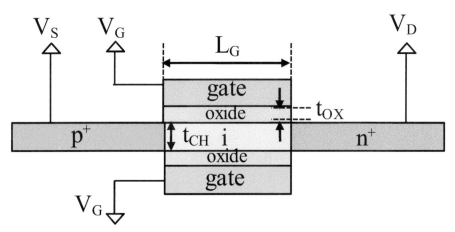

**Figure 3.5**   Structure of simulated n-type double-gate Tunnel FET.

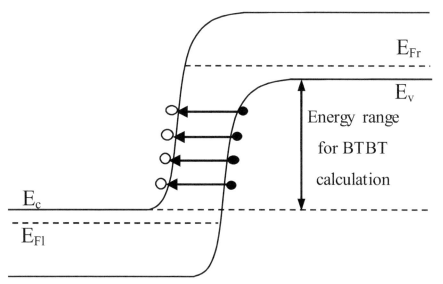

**Figure 3.6**   Schematic of non-local BTBT calculation range. Adapted from [5].

$E_{Fl}$ is the electron quasi-Fermi level and $E_{Fr}$ the hole quasi-Fermi level. $T(E)$ is the tunneling probability of the carrier calculated using the Wentzel-Kramers-Brilloui (WKB) approximation and given by Equation (3.6). The quantum wave vector $k(x)$ is calculated according to individual wave vectors of hole and electron as expressed by Equation (3.7).

$$J(E)\,\Delta E = \frac{qKT\sqrt{m_e m_h}}{2\pi^2\hbar^3}T(E)$$

$$\log\left(\frac{\left(1 + exp\left[\frac{E_{Fr}-E}{k_B T}\right]\right)\left(1 + exp\left[\frac{E_{Fl}-E-E_{max}}{k_B T}\right]\right)}{\left(1 + exp\left[\frac{E_{Fl}-E}{k_B T}\right]\right)\left(1 + exp\left[\frac{E_{fr}-E-E_{max}}{k_B T}\right]\right)}\right)\Delta E \quad (3.5)$$

$$T(E) = exp\left(-2\int_{xstart}^{xend} k(x)\,dx\right)$$

$$\approx exp\left[-\frac{4\sqrt{2m^*}\,E_g^{3/2}}{3\hbar qF}\right]$$

$$\approx exp\left[-\frac{4\sqrt{2m^*}\,E_g^{3/2}}{3\hbar q\,(\Delta\Phi + E_G)}\sqrt{\frac{\varepsilon_{si}}{\varepsilon_{ox}}}t_{si}t_{ox}\right]\Delta\Phi \quad (3.6)$$

$$k\left(x\right) = \frac{k_e k_h}{\sqrt{k_e^2 + k_h^2}}, \quad k_e = \frac{1}{i\hbar}\sqrt{2m_0 m_e(x)(E - E_c(x))},$$

$$k_h = \frac{1}{i\hbar}\sqrt{2m_0 m_h(x)(E_v - E))} \tag{3.7}$$

### 3.2.2 Dielectric Permittivity, EOT, and Body Thickness Impact

The first physical parameter evaluated is the relative permittivity of the oxide material between the gates and channel. The source, channel, and drain regions of the n-type TFET are simulated as Si ($E_G$ = 1.12 eV). The p-type source doping is simulated with a concentration of $1 \times 10^{20}$ atoms/cm$^3$, the n-type channel with $1 \times 10^{17}$ atoms/cm$^3$ and n-type drain with $5 \times 10^{18}$ atoms/cm$^3$. Doping concentrations in the same range were used by several TFET-based works [6–8]. The gate length is simulated with 20 nm. Both the body and oxide thicknesses are simulated with 5 nm. The gate work function in all the simulations was chosen such that the conduction band in the channel region is aligned to the Fermi level of the source region. This allows BTBT generation for a $V_{GS}$> 0 V. Traps at the silicon/oxide interface are not included.

As shown in Figure 3.7, oxides with large relative permittivity $\varepsilon_r$ (or high-k) allow for better electrostatic control of the gate over the channel. Consequently, a large tunneling current is possible with a lower potential difference between the gate and the source regions.

As the SS behavior of TFETs is different from MOSFETs, i.e. the SS changes with $V_{GS}$ magnitude in contrast to the constant SS value of MOSFETs in the sub-threshold region, there are two parameters of interest to be extracted from the input characteristics of TFETs: the point slope (or minimum SS) which is defined as the minimum swing value at any point of the I–V characteristic; and the average slope which is calculated between the voltage at which the current starts to increase ($V_{OFF}$) and the threshold voltage (here defined as $|V_{GS}|$ applied for achieving a current level of 0.1 µA/µm). The average SS can be calculated as expressed by Equation (3.8) [3]:

$$SS_{AV} = \frac{V_{TH} - V_{OFF}}{log\left(0.1\,\mu\,\left[\frac{A}{\mu m}\right]\right) - log\left(I_{OFF}\,\left[\frac{A}{\mu m}\right]\right)} \tag{3.8}$$

As presented in Figure 3.7 and in comparison to SiO$_2$, the use of materials with high-k (HfO$_2$ and ZrO$_2$) increases the leakage current of the tunneling device. Despite such increase, the average SS is shown to

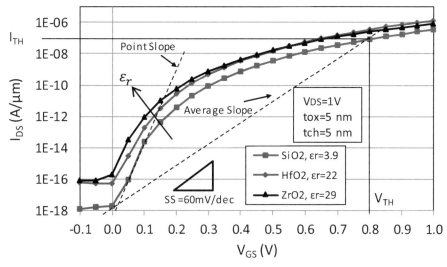

**Figure 3.7** Impact of oxide material on the input characteristics of the Si-TFET at room temperature.

decrease: $SS_{AV} \approx 72$ mV/dec $(SiO_2)$, $SS_{AV} \approx 68$ mV/dec $(ZrO_2)$ and $SS_{AV} \approx 64$ mV/dec $(HfO_2)$. Despite the improvements in the average slope, the minimum slope with high-k materials is shown degraded: 21 mV/dec $(SiO_2)$, 23 mV/dec $(ZrO_2)$ and 32 mV/dec $(HfO_2)$.

In Figure 3.8, the decrease of the equivalent oxide thickness (EOT) (considering $HfO_2$ as high-K material) is shown to increase the drive-current, improving the average slope of the I–V characteristics and maintaining the leakage current below fA/$\mu$m. The EOT is calculated as follows:

$$EOT = t_{highK}\left(\frac{\varepsilon_r \; SiO_2}{\varepsilon_r \; high_k}\right) \tag{3.9}$$

As expressed by Equation (3.6), the increase of drive-current with lower EOT values is resultant from the increased tunneling transmission probability due to the increase of oxide capacitance given by $C_{ox} = \varepsilon_{ox}/t_{ox}$.

Maintaining a $HfO_2$ oxide with a thickness of 2.5 nm, and the same doping concentration in the TFET regions, a change in the silicon body thickness $t_{OX}$ was performed with the results presented in Figure 3.9. According to the tunneling transmission probability expressed by Equation (3.6), reducing the body thickness of the device allows for better electrostatic control of the gate over the channel, and consequently a larger tunneling current.

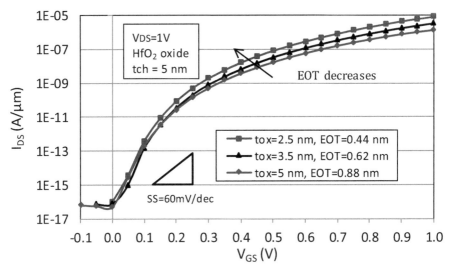

**Figure 3.8**　Impact of oxide thickness (considering $HfO_2$) on the input characteristics of the Si-TFET at room temperature.

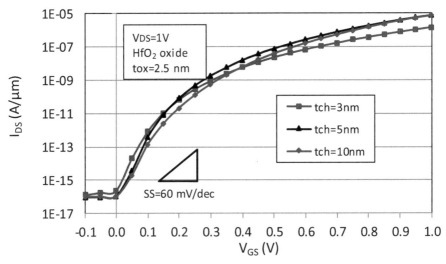

**Figure 3.9**　Impact of body thickness on the input characteristics of the Si-TFET at room temperature.

This is observed in Figure 3.9 for $V_{GS}$ values below 0.15 V. At larger $V_{GS}$ magnitudes this tendency is not observed.

In the same figure it is shown that reducing the channel thickness from 10 nm to 5 nm increases the tunneling current over the entire $V_{GS}$ range considered, but a further reduction attenuates the current. This behavior can be explained due to the reduced cross-sectional area (with $t_{CH} = 3$ nm) available for current to flow, showing that there is a body thickness that maximizes the performance of the double-gate device [3]. The results also show that decreasing the body thickness of the device has a minimum effect on the leakage current. In the following section, the double-gate TFET is simulated considering an $HfO_2$ oxide with a thickness of 2.5 nm and a body thickness of 5 nm.

### 3.2.3 Impact of Doping in Drain and Source Regions of Si-TFET

In TFETs, the doping levels in the source and drain regions must be carefully optimized in order to improve the electrical characteristics of the device, i.e. improve the on-current and decrease the leakage current. In Figure 3.10(a), the impact of drain doping concentration (with abrupt profile) of an n-type double-gate Si-TFET in the input current characteristics is presented. The source doping is considered as $N_A = 1 \times 10^{20}$ atoms/cm$^3$. It is shown that an equal source and drain doping concentration results in a tunneling device with ambipolar behavior with negative $V_{GS}$ values.

With a negative $V_{GS}$ and large doping concentration in the drain region, the energy bands in the channel region bend up and a tunneling conduction predominant by holes (hole-BTBT) is enabled at the drain-channel interface. As shown in Figure 3.10(b), the decrease of the drain doping concentration

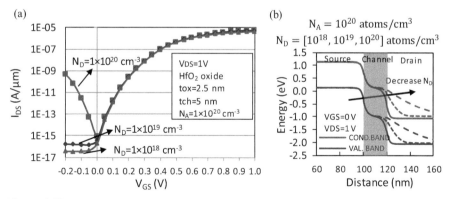

**Figure 3.10** (a) Impact of drain doping on the input characteristics of the Si-TFET and (b) respective energy band diagram.

increases the conduction and valence bands in the drain region, minimizing the hole-BTBT mechanism (considering negative $V_{GS}$), and the leakage current of the device (at $V_{GS} = 0$ V). A low drain doping concentration is therefore required to attenuate the ambipolar current of TFETs.

As shown in Figure 3.11, the on-current of TFETs is highly dependent on the source-doping concentration, since tunneling takes place between the source and channel regions. In the simulations, a low drain doping concentration of $N_D = 1 \times 10^{18}$ atoms/cm$^3$ is considered. As shown in Figure 3.12, the increase of source-doping concentration results in increased energy bands in the source region, therefore reducing the tunneling barrier for electrons to tunnel to the drain side when the device is operating in the on-state. A steeper transition between the energies of the source and channel regions also increases the magnitude of the electric field $F$ applied between the regions, therefore increasing the tunneling probability as expressed by Equation (3.6).

A direct consequence of a large source-doping concentration ($N_A = 1 \times 10^{21}$ atoms/cm$^3$) is the increase of the leakage current. For this case and considering a $V_{GS} = 0$ V, there is an increased probability of tunneling due to the reduction of the tunneling barrier at the source-channel regions and a consequent increase of BTBT current.

The doping type profile (abrupt vs. gradient) of *TFETs* is also an important factor to take into account. As shown by several works, a low

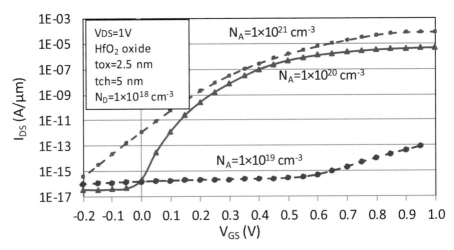

**Figure 3.11**    Impact of source doping on the input characteristics of the Si-TFET at room temperature.

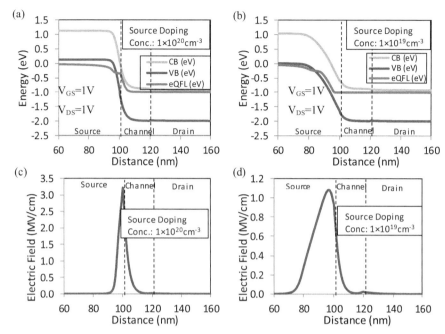

**Figure 3.12** Impact of source-doping concentration on the energy bandgap (a) and (b) and electric field (c) and (d) considering $N_A = 1 \times 10^{20}$ cm$^{-3}$ and $N_A = 1 \times 10^{19}$ cm$^{-3}$ and the TFET device in the on-state ($V_{DS} = V_{GS} = 1$ V).

source-doping gradient is required in order to avoid a gradual band bending in the source-channel interface and a consequent decrease of BTBT current due to a lower electric-field magnitude applied in the regions of the device [6–8]. In [6], the authors have shown that the SS of a Si-based TFET increases from 75 to 115 mV/dec with increased source gradient. The effect of degraded SS with the increase of source-doping gradient is shown to be more important in short-channel (with $L_G < 50$ nm) TFETs due to the increase of leakage current resultant from the increase of direct tunneling between the source and drain regions.

In [7], the authors have shown by simulations that a Gaussian drain doping profile can improve the electrical characteristics of the TFET due to the increase of the drain-channel depletion width, and the consequent reduction of leakage current. As the gate-to-drain parasitic capacitance ($C_{GD}$) of TFETs is reduced with large drain doping gradients, several *RF* parameters such as cutoff frequency $f_T$, maximum oscillation frequency $f_{MAX}$, and gain bandwidth *GBW* can benefit from the Gaussian drain doping profile.

In Figure 3.13, the impact of source-doping concentration (considering uniform doping profile) in the output characteristics of the double gate Si-n-TFET under study is presented, considering a $V_{GS}$ = 0.5 V. It is shown that the increase of the source-doping concentration results not only in larger on-current values but also saturated currents at lower drain bias. In the simulations, $V_{DS}$-saturation is considered at the drain-to-source voltage where the current achieves 90% of its saturated value.

As opposed to MOSFETs, in TFETs the current presents an exponential increase at low $V_{DS}$ and a saturated behavior at large drain bias. Under low $V_{DS}$ the TFET channel is inverted due to the injection of electrons from the drain region, thus creating a *p-n* junction like built-in potential.

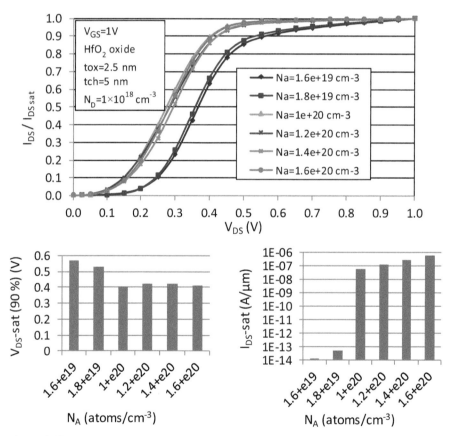

**Figure 3.13**   Impact of source doping in the output characteristics of the Si-TFET at room temperature.

This potential can be controlled by the source-doping concentration and by the $V_{GS}$ magnitude [8]. At large drain bias, the BTBT current starts to saturate due to the saturation of the channel conduction band, whose energy cannot be further decreased by larger $V_{DS}$ values. Therefore, the energy window $\Delta\Phi$ cannot increase with further $V_{DS}$ values, and the BTBT current is limited.

### 3.2.4 Impact of Materials in a Double-gate TFET

As shown by the previous results, BTBT currents of some $\mu A/\mu m$ in Si-based TFETs are possible at large voltage, i.e. $V_{GS} = V_{DS} = 1$ V (see Figure 3.10). Although the current magnitude of Si-TFETs is very low for considering TFETs as a candidate for MOSFET replacement, it can be enough for a wide range of low-power applications. Against Si-based TFETs is the fact that due to the low energy bandgap of silicon ($E_G = 1.12$ eV), large $V_{GS}$ values (electric field $F$ increases with $V_{GS}$) are required for increased tunneling transmission probability $T(E)$ (see Equation (3.6)). In order to achieve a similar $T(E)$ at lower $V_{GS}$ values, TFETs designed with lower energy bandgaps than that of the silicon are required.

A comparison of three double-gate TFETs designed with different materials (Si, Ge and InGaAs) is shown in Figure 3.14. Compared to silicon,

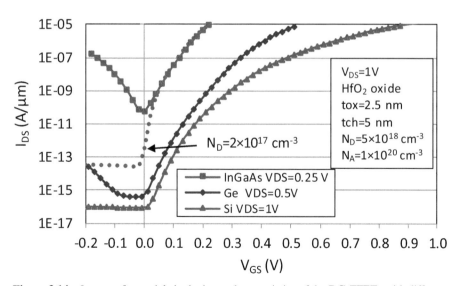

**Figure 3.14** Impact of materials in the input characteristics of the DG-TFETs with different materials at room temperature.

the use of germanium with $E_G = 0.66$ eV or InGaAs with $E_G = 0.571$ eV allows for larger BTBT currents at lower $V_{GS}$ values due to their increased transmission probability [2]. As mentioned in Chapter 2, the implementation of group III–V materials in TFETs is under great research due to the low energy bandgap and mass values compared to group IV materials such as silicon and germanium.

As shown in Figure 3.14, the leakage current of TFETs increases with lower $E_G$ materials. Physical effects such as Shockley–Read–Hall generation in the source ($p^+$) and drain ($n^+$) regions (as shown in Figure 3.4) are responsible for such increase and consequent degradation of SS. Ambipolarity is also observed for the InGaAs with a source and drain doping concentration of respectively $N_A = 1 \times 10^{20} \text{cm}^{-3}$ and $N_D = 5 \times 10^{18} \text{cm}^{-3}$. As previously mentioned, the reduction of drain doping concentration can attenuate the leakage current and the ambipolarity effect as shown by the dashed red curve of Figure 3.14.

### 3.2.5 Impact of Doping in Drain and Source Regions for TFETs with Different Materials

In this section, the impact of source and drain doping concentrations (considering uniform doping profile) in double-gate n-type TFETs designed with different materials is presented. For simulation purposes, the TFETs present an HfO$_2$ oxide with 2.5 nm and body thickness of 5 nm. In Figure 3.15, the impact of the source-doping concentration is analyzed, considering a fixed drain doping concentration of $N_D = 1 \times 10^{18} \text{cm}^{-3}$. The $V_{GS}$ magnitude is considered to be twice that of $V_{DS}$. Such a diode-connection type is well known and used in several circuit applications such as rectifiers and charge-pumps. In these circuits, the transistor has to present the lowest possible internal resistance when forward biased and the largest under reverse-bias conditions. As TFETs present a different doping structure than that of MOSFETs, a study on the impact of the source and drain doping concentration in the internal resistance of TFETs with different materials and diode-connection type is required.

In Section 3.2.3 it was observed that in a forward biased Si-TFET ($V_{DS} > 0$ V) the BTBT current increases with the increase of the source-doping concentration due to the larger electric field applied between the source-channel regions interface. As shown in Figure 3.15, the same behavior is observed considering TFETs designed with lower energy bandgap materials such as Ge and InGaAs. The trend increase of the leakage current at large

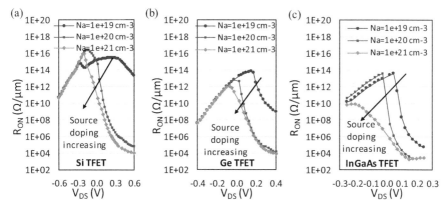

**Figure 3.15** Impact of source-doping concentration for different TFET materials in the internal resistance of the device, considering $V_{GS} = 2V_{DS}$ and a drain doping concentration of $N_D = 1 \times 10^{18}$ cm$^{-3}$.

source-doping concentrations is observed for Ge and InGaAs-based TFETs with the latter showing larger leakage due to the closer proximity between its conduction and valence band.

Figure 3.16 presents the performance of the three mentioned diode-connected TFETs considering changes in the drain doping. As previously explained, the drain-doping concentration does not affect the BTBT current when the device is largely forward biased. This is because tunneling takes place between the source and channel interface. In Section 3.2.3, it was

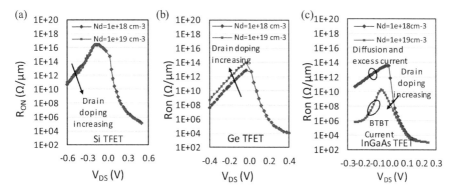

**Figure 3.16** Impact of drain-doping concentration for different TFET materials in the internal resistance of the device, considering $V_{GS} = 2V_{DS}$ and a source-doping concentration of $N_A = 1 \times 10^{20}$ cm$^{-3}$.

shown that for a Si-TFET the reduction of the drain doping concentration can reduce the ambipolarity of the device and also the leakage current. As shown in Figure 3.16(a), when the Si-TFET is reverse-biased ($V_{DS} < 0$ V) the intrinsic *p-i-n* diode of the TFET is active (forward biased) and the carrier injection mechanism is characterized by diffusion and excess current. No BTBT current is observed in this case. It is also shown that at large reverse bias, the internal resistance of the Si-TFET decreases due to the increase of reverse current. For a Si-TFET with diode-type connection, changing the drain doping has a minimum effect in the reverse electrical characteristics.

In contrast, for a Ge-TFET the increase of the drain-doping concentration is shown to decrease the reverse current of the device (Figure 3.16(b)). However, this is not the case for TFETs designed with InGaAs materials as shown in Figure 3.16(c). In this case, for a reverse-biased TFET, the increase of the drain-doping result in an unwanted increase of the reverse current and consequent decrease of the internal resistance of the device. The different behavior between Ge-based and InGaAs-based TFETs is explained by the energy band diagrams shown in Figure 3.17.

As shown in Figure 3.17(a) and considering a $V_{DS} = -0.25$ V and $V_{GS} = -0.5$ V, the increase of drain-doping concentration in the Ge-TFET increases the energy barrier between the channel and drain regions, with a consequent reduction of diffusion and excess current. In contrast if the energy bandgap is very low, the increase of the drain-doping concentration will not only increase the energy barrier between the drain and channel regions but also enable a

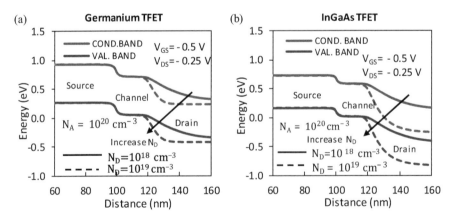

**Figure 3.17**    Energy band diagram for (a) Ge-TFET and (b) InGaAs-TFET for a reverse-biased TFET.

reverse BTBT current between the same regions. As shown in Figure 3.17(b), a large increase in the drain doping concentration locates the conduction band of the drain region below the valence band of the channel, enabling a tunneling probability of holes to tunnel to the source region. Therefore, in TFETs, the choice of drain doping concentration has to be carefully chosen in order to mitigate the diffusion and excess current, without enabling a reverse BTBT mechanism between the drain and source regions.

## 3.3 Chapter Summary

In this chapter, simulations at a device level show that the performance of TFETs is strongly dependent on several physical parameters. The chapter conclusions are shown as follows:

- Gate oxides with large permittivity are shown to improve the gate control over the channel with a consequent increase of the electrical field between the junctions. This increases the BTBT probability, thus increasing the drive-current. The consequent increase of the leakage current with large oxide permittivity degrades the minimum SS (or point slope) of the device contrasting with the improved average slope;
- The decrease of the equivalent oxide thickness is shown to increase the drive-current of the TFET and to improve the average slope of the I–V characteristics with no major effects in the leakage current;
- There is a channel thickness that maximizes the drive current of the TFET. Reducing the channel thickness of the TFET results in increased tunneling probability and hence increased BTBT current. However, the decrease of the channel thickness below some dimensions results in a reduced cross-sectional area available for current to flow, with a consequent attenuation of drive current. The simulated results show that decreasing the body thickness of the device has a minimum effect on the leakage current;
- In TFETs, equal source and drain-doping concentrations characterizes the device with ambipolarity behavior. With a negative $V_{GS}$ (in n-TFET) and large doping concentration in the drain region, the energy bands in the channel region bend up and a tunneling conduction predominant by holes (hole-BTBT) is enabled at the drain-channel interface. Therefore, in TFETs, the choice of the drain-doping concentration has to be carefully chosen in order to mitigate the diffusion and excess current,

without enabling a reverse BTBT mechanism between the drain–channel interface;

- The increase of source-doping concentration allows for an increased electrical field magnitude applied between the source-channel regions and therefore increased BTBT current with a consequent increase of the leakage current and SS;
- Compared to silicon-based TFETs, the use of lower energy bandgap materials (Ge or InAs) is shown to improve the device performance at lower gate voltage magnitudes with a consequent degradation of the leakage current. This behavior is directly related to the decrease of barrier width between the source-channel regions.

## References

[1] Ionescu, A. M., and Riel, H. "Tunnel field-effect transistors as energy-efficient electronic switches," *Nature* 479, 329–337, 2011.

[2] Seabaugh, A. C., and Zhang, Q. "Low-voltage tunnel transistors for beyond CMOS logic," in *Proceedings of the IEEE*, vol. 98, no. 12, pp. 2095–2110, 2010.

[3] Boucart, K., and Ionescu, A. M. "Double-gate tunnel FET with high-k gate dielectric," in *IEEE Trans. Electron Devices*, vol. 54, no. 7, pp. 1725–1733, 2007.

[4] Knoch, J., Mantl, S., and Appenzeller, J. "Impact of the dimensionality on the performance of tunneling FETs: Bulk versus one-dimensional devices," *Solid-State Electr.* 51, 572–578, 2007.

[5] Atlas User's Manual, Silvaco, Inc., Santa Clara, CA, United States, 2014.

[6] Chen, S., Huang, Q., and Huang, R. "Source doping profile design for Si and Ge tunnel FET," in *ECS Transactions*, vol. 60, no. 1, pp. 91–96, 2014.

[7] Vijayvargiya, V., and Kumar, S. "Effect of drain doping profile on double-gate tunnel field-effect transistor and its influence on device RF performance," in *IEEE Transactions on Nanotechnology*, vol. 13, no. 5, 2014.

[8] Rajamohanan, B., Mohata, D., Ali, A., and Datta, S. "Insight into the output characteristics of III–V tunneling field effect transistors," *Appl. Phys. Lett.* 102:092105, 2013.

# 4

# Tunnel FET: Electrical Properties

In this chapter, the electrical properties of tunnel field-effect transistors (TFETs) are explored for digital and analog design. In order to perform simulations at device and circuit level, models that describe the electrical characteristics of TFETs are required. In the following sections, TFET models from the literature based on (1) analytic equations and (2) lookup tables (LUTs) are described. With a focus on the Verilog-A-based LUT-TFET models, a comparison between digital and analog figures of merit (FOM) is extracted and compared to those of conventional thermionic metal-oxide-semiconductor field-effect transistors (MOSFETs).

## 4.1 Tunnel FET Models for SPICE Simulations

There is currently a great research effort in order to model the static and dynamic behavior of TFETs for circuit simulation purposes. The TFET current has to be calculated in the four quadrants of operation presented in Figure 4.1. In this section, two distinct TFET models from the literature are described: one based on physical equations and another based on LUTs describing the main electrical characteristics of the device.

### 4.1.1 Analytic TFET Model

In order to make performance projections of TFETs at device and circuit level, a universal SPICE model for an n-type TFET was developed at the University of Notre Dame, USA, capturing the essential features of the tunneling process in both quadrants of operation presented in Figure 4.1 [1–3]. Based on the Kane–Sze formula for Zener tunneling, the model captures the bias-dependent inverse subthreshold slope (SS) in the input characteristics $I_{DS}$–$V_{GS}$, the super-linear drain current onset in the output

| Q2 | $V_{DS}$ | Q1 |
|---|---|---|
| $V_{GS} < 0$ V | | $V_{GS} > 0$ V |
| $V_{DS} > 0$ V | | $V_{DS} > 0$ V |
| Ambipolar | | Kane-Sze |
| Q3 | | Q4 $V_{GS}$ |
| $V_{GS} < 0$ V | | $V_{GS} > 0$ V |
| $V_{DS} < 0$ V | | $V_{DS} < 0$ V |
| Diode | | NDR |

**Figure 4.1**   Regions of operation in n-TFET.

characteristics $I_{DS}$–$V_{DS}$, the ambipolarity conduction at negative $V_{GS}$, and the Esaki tunnel current with negative differential resistance (NDR) at reverse bias (negative $V_{DS}$). The analytic model was developed and validated with a double-gate (DG) InAs-TFET with the gate perpendicular to the tunnel junction [4] and with atomistic simulations of a broken-gap AlGaSb-InAs TFET with the gate parallel to the tunnel junction [5].

The equivalent circuit of the TFET at a device level is shown in Figure 4.2 and includes a voltage-controlled current source $I_{DS}$, intrinsic capacitances $C_{GS}$ and $C_{GD}$, and gate/source/drain series resistors $R_G$, $R_S$, and $R_D$, respectively.

In Figure 4.3, the TFET model is fitted to two different TFET structures. The first is a DG InAs-TFET [4] with the characteristics extracted by an atomistic quantum-mechanical device simulator, while the second is a single-gate broken-gap AlGaSb/InAs-TFET [5] with the characteristics predicted by Synopsis technology computer-aided design (TCAD). Both TFETs are

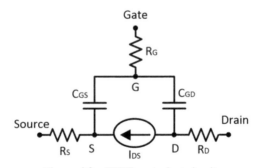

**Figure 4.2**   TFET equivalent circuit.

*Source*: Adapted from [1].

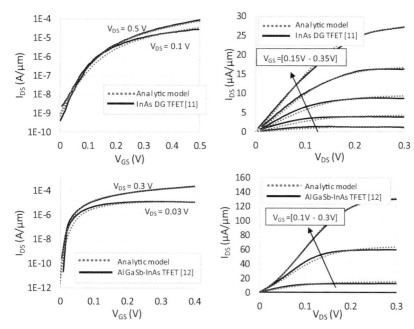

**Figure 4.3** Comparison between modeled and simulated input and output characteristics of a double-gate InAs tunnel field-effect transistor (TFET) [4] and broken gap AlGaSb/InAs TFET [5].

designed with a channel length of 20 nm. As shown by the results, the analytic model agrees well with the simulated results, capturing the main characteristic of TFETs: low leakage current and steep *SS* in the input characteristics $(I_{DS}-V_{GS})$ and exponential increase followed by linear onset in the output characteristics $(I_{DS}-V_{DS})$.

Despite similar results between the TFET models (analytic and simulated) under forward bias conditions, further validation is required under the opposite state, i.e., reverse bias condition (intrinsic p-i-n diode of the TFET is forward biased). This validation is important in order to evaluate the performance of TFETs in several circuit applications where the device is not only subjected to forward bias conditions but also reverse bias. A model describing the main electrical characteristics of p-type TFETs (*I–V* and *C–V*) is also of interest for the projection of TFET-based circuits. In the following subchapter, models based on LUTs describing the main electric characteristics of both n-type and p-type TFETs are presented.

## 4.1.2 TFET Model Based on Lookup Tables

In this section, two TFET models from the literature and based on LUTs are presented. The TFETs were designed with group III–V materials and different structure configurations: a double-gate (DG) InAs homojunction TFET and a DG GaSb-InAs near broken gap heterojunction TFET. Current–voltage ($I$–$V$) and capacitance–voltage ($C$–$V$) characteristics were obtained from the TCAD Sentaurus device simulator by the NDCL group of the PennState University [6]. The calibrated TFET models serve as an approximation of full-band atomistic calculation of TFET band diagram and BTBT current to generate the DC characteristics. The intrinsic capacitances: gate-to-drain $C_{GD}$ and gate-to-source $C_{GS}$ were obtained from the *TCAD* small-signal simulations and validated with measured transient characteristics of TFETs [7, 8]. In order to explore the potential energy efficiency benefits of TFETs for low power design, the LUT-TFET models have been widely applied in several works and areas such as SRAM/digital [9–12] and ultralow power analog design [13–15]. The physical characteristics of the TFETs are presented in Table 4.1.

Contrasting with the analytic TFET model described in the previous section, the LUT-TFET models are able to characterize the TFET device current in the four quadrants of operation shown in Figure 4.1 for n- and p-type structures. In addition and contrasting with the analytic model, the intrinsic capacitances $C_{GS}$ and $C_{GD}$ of the LUT-TFETs are modeled and shown dependent on the $V_{DS}$ and $V_{GS}$ bias. Therefore, the LUT-TFET models are more suitable to be applied in circuit design and simulation for performance projections in comparison to the analytic model previously described.

**Table 4.1**  Physical parameters of the double-gate tunnel field-effect transistors modeled by the lookup tables

| Structure | DG InAs-TFET | DG-GaSb/InAs TFET |
| --- | --- | --- |
| Source material | InAs | GaSb |
| Source doping concentration | $4 \times 10^{19} \mathrm{cm}^{-3}$ | $4 \times 10^{19} \mathrm{cm}^{-3}$ |
| Channel material | InAs | InAs |
| Drain material | InAs | InAs |
| Drain doping concentration | $6 \times 10^{17} \mathrm{cm}^{-3}$ | $2 \times 10^{17} \mathrm{cm}^{-3}$ |
| Channel length | 20 nm | 40 nm |
| Channel thickness | 5 nm | 5 nm |
| Oxide material | $HfO_2$ | $HfO_2$ |
| Oxide thickness | 5 nm | 2.5 nm |

In the following sections, the electric characteristics of the DG III–V TFET devices described in Table 4.1 (LUT-based) are explored for digital and analog design. The study will allow to identify the voltage range where the TFET technology can outperform conventional MOSFETs at an energy efficiency level.

## 4.2 Electrical Characteristics of TFETs

In this section, a comparison between the input and output characteristics of TFETs and conventional complementary metal-oxide-semiconductor (CMOS) devices is performed in order to explore the voltage range where the former outperforms the latter. A thermionic MOSFET device is simulated as a triple-gate structure (FinFET) with a gate length of 20 nm, a fin height ($F_H$) of 28 nm, and a fin width ($F_W$) of 15 nm. FinFET simulations are based on a predictive technology model from the NIMO group of the Arizona State University [16]. The TFET simulations are based on the LUTs previously described: a DG InAS-TFET (homojunction) and a DG-GaSb-InAs TFET (heterojunction).

### 4.2.1 Input Characteristics of TFETs

In Figure 4.4, the input characteristics of n-type (a) and p-type (b) FinFETs are presented. The channel width is calculated as follows: $W_{FIN} = N_{FINS} \times (2 \times F_H + F_W)$. In order to calculate the current for a channel width of 1 µm, 14 Fins ($N_{FINS}$) are considered. When considering a gate length of 20 nm and a $|V_{DS}|$ range of 0.1–0.5 V, the leakage current of FinFETs ($I_{DS}$ at $V_{GS} = 0$ V) is shown to be larger than 10 nA/µm. It is also shown that the SS of FinFETs is constant over a wide range of $V_{GS}$ in both n- and p-types [−0.4 to 0.4 V]. As explained in the previous chapter and according to the equation expressed by (3.1), the SS value of *MOSFETs* in the subthreshold region is directly related to temperature, rather than the $V_{GS}$ magnitude.

In contrast, Figures 4.5 (homojunction TFET) and 4.6 (heterojunction TFET) show that the SS of TFETs is dependent on the $V_{GS}$ magnitude applied to the device (SS not constant). Considering Equation (4.1) as the BTBT current [17], the SS of TFETs can be expressed according to Equation (4.2) [18]:

$$I_{BTBT} = a.F.V_R.e^{-b/F} \tag{4.1}$$

$$SS = \left( \frac{\partial log(I_D)}{\partial V_{GS}} \right)^{-1} = ln(10) \times \left[ \frac{1}{V_R} \frac{\partial V_R}{\partial V_{GS}} + \frac{\partial F}{\partial V_{GS}} \left( \frac{F+b}{F^2} \right) \right]^{-1} \tag{4.2}$$

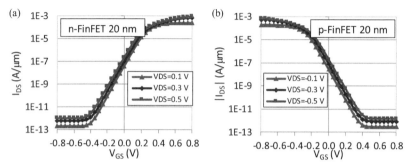

**Figure 4.4**    Input characteristics of (a) n-type FinFET and (b) p-type FinFET configurations.

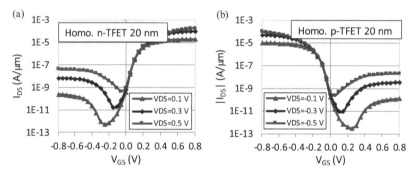

**Figure 4.5**    Input characteristics of double-gate InAs-based tunnel field-effect transistor (homojunction) for (a) n-type and (b) p-type configurations.

In Equation (4.1), the BTBT current shows a dependence on the reverse bias $V_R$ ($V_{GS}$ and $V_{DS}$ dependent, see [1]), electric field magnitude $F$ applied on the tunnel junction, and two terms $a$ and $b$ that are material dependent. As shown in Equation (4.2), there are two terms (not thermal dependent) that have to be maximized in order to minimize the slope of current–voltage characteristics of TFETs.

The first term indicates that the transistor has to be engineered so that $V_{GS}$ directly controls the tunnel junction bias $V_R$, assuring that the gate field directly modulates the channel. This can be achieved with a tunneling device with a high-k gate dielectric and ultrathin body. In a tunneling device with a channel well controlled by the gate, the term $\partial V_R / \partial V_{GS}$ is approximated as 1 [18]. As $V_R$ is directly related to $V_{GS}$ (see [1]), the SS of a TFET increases with $V_{GS}$.

The second term of the denominator in Equation (4.2) dictates that in order to minimize SS, the derivative of the junction electric field in function

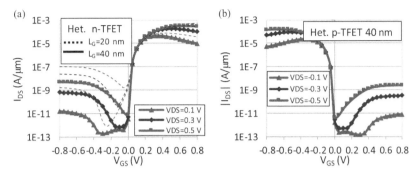

**Figure 4.6** Input characteristics of double-gate GaSb-InAs-based tunnel field-effect transistor (heterojunction) for (a) n-type and (b) p-type configurations.

of $V_{GS}$ should be maximized. In TFETs, the increase of gate bias reduces the tunneling width with a consequent increase of the junction electric field. This indicates that by increasing $V_{GS}$ magnitude, a lower SS is achieved. However, the decrease of the tunneling width does not present a linear relation with $V_{GS}$, and at some gate bias, the tunneling width of the tunneling device cannot decrease anymore (as also the junction electric field). This will decrease the derivative of the electric field and consequently increase the SS of the tunneling device.

As shown by the results presented in Figures 4.5 and 4.6, under negative $V_{GS}$ (considering n-TFETs) and positive $V_{GS}$ (considering p-TFETs), the current magnitude increases due to the ambipolar nature of TFETs. In contrast, the reverse current of the Si-FinFET is shown to decrease with the increase of reverse gate bias. One can also observe that the current of TFETs at large gate bias is lower than that of the Si-FinFET at similar drain bias magnitudes.

Due to the lower tunneling transmission probability consequent of lower energy bandgap materials, the homojunction TFET is shown to produce lower drain current magnitudes when compared to the heterojunction counterpart. As the drain regions of both homojunction and heterojunction TFETs are designed with the same material (InAs) and similar doping concentrations, both TFETs present a similar leakage current for a gate length $L_G$ of 20 nm. However, the increase of the channel length in the heterojunction TFET to 40 nm is shown to decrease the leakage current by at least two orders of magnitude, while maintaining a larger drain current magnitude when compared to the homojunction counterpart. The decrease of the leakage current with the increase of channel length can be explained by the reduction of short-channel effects described in Section 4.3.

## 4.2.2 Output Characteristics of TFETs

A comparison between the output characteristics of Si-FinFET, heterojunction TFET ($L_G$ = 40 nm), and homojunction TFET ($L_G$ = 20 nm) is shown in Figure 4.7 considering two distinct bias conditions: a low $V_{GS}$ = 0.2 V and $V_{GS}$ = 0.5 V. With both conditions, one can observe that for a short channel length of 20 nm, the drive current of the Si-FinFET is dependent on the drain voltage due to the channel length modulation effect. Another characteristic of conventional thermionic devices is the bidirectional conduction at both positive and negative drain bias resultant from the similar doping types of both source and drain regions.

In contrast, TFETs present particular electrical characteristic due to the different doping structure and carrier injection mechanism. With a reverse drain bias and considering a moderate $V_{GS}$ magnitude, an NDR region can be observed in both n- and p-type TFETs. With a negative (positive) $V_{DS}$, the intrinsic p-i-n structure of the n (p)-TFET is forward biased and the reverse current follows the characteristic of the Esaki tunnel diode. In Figure 4.8, the magnitude of current at reverse drain bias is shown for (a) heterojunction n-TFET and (b) homojunction n-TFET. At large reverse drain bias, the current of TFETs increase due to the increase of excess and diffusion current as explained in Section 4.2.1.

The particular carrier injection mechanism of TFETs characterizes the forward current with large saturation at large drain bias and a threshold voltage (denominated here as drain threshold voltage $V_{THD}$) with a clear dependence on the gate bias. In TFETs, a minimum value of drain voltage is required in order to turn the device on. This characteristic is independent on the gate bias applied [1, 19]. The reason of such behavior is that the decrease of the energy barrier between regions is calculated as a function of both gate and drain bias applied to the device.

In [19], the authors define the drain threshold voltage as the drain bias for which the drain current dependence changes from a quasi-exponential to a linear behavior. In Figure 4.7(b), one can observe that at low drain bias ($V_{DS}$ < 0.4 V) and considering $V_{GS}$ = 0.2 V, the heterojunction TFET presents a higher drive current in comparison to the homojunction counterpart and the thermionic FinFET. This indicates that the TFET technology can surpass the performance of thermionic devices at low bias conditions. In contrast, at large bias the Si-FinFET presents a higher drive current as shown in Figure 4.7(a).

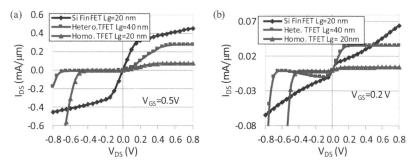

**Figure 4.7** Comparison of output characteristics for n-type devices considering (a) $V_{GS} = 0.5$ V and (b) $V_{GS} = 0.2$ V.

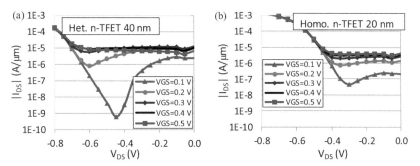

**Figure 4.8** Negative differential resistance at negative $V_{DS}$ for (a) heterojunction n-tunnel field-effect transistor (TFET) and (b) homojunction n-TFET.

## 4.2.3 Intrinsic Capacitance of TFETs

In TFET devices, the intrinsic gate-to-source $C_{GS}$ and gate-to-drain $C_{GD}$ capacitances present a different characteristic in function of $V_{GS}$ when compared to conventional thermionic MOSFETs. In TFETs, the total gate capacitance is mostly entirely reflected by $C_{GD}$. As the TFET current is dependent on the shrinking barrier in the source–channel interface, the resultant $C_{GS}$ is very low during the on-state of the transistor. The large $C_{GD}$ of TFETs occurs due to the low potential drop between the channel and the drain when the transistor operates in on-state [7]. The TFET $C_{GD}$ and $C_{GS}$ characteristics are presented in Figures 4.9(a, b) for homojunction and heterojunction TFETs and for n- and p-type configurations.

Another signature of TFET devices is the presence of the "pinchoff" point in $C_{GD}$ that occurs at increased $V_{GS}$ bias with increased $V_{DS}$ values. This particular behavior is due to the larger energy band bending at the

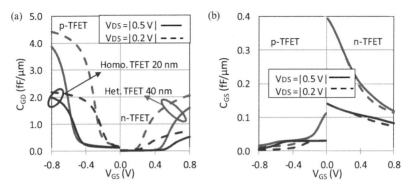

**Figure 4.9**    Gate-to-drain (a) and gate-to-source (b) intrinsic capacitances of homojunction and heterojunction tunnel field-effect transistors.

source–channel interface with larger $V_{GS}$ magnitudes and the consequent change of the potential drop between the channel and the drain regions [8].

## 4.3 TFETs in Digital Design

Several works have explored the performance of TFETs for ultralow power applications. As an example, in [20], the performance of a SiGe-TFET was compared to that of a SOI-MOSFET at a device level. The authors concluded that TFETs can achieve a 10× speed improvement at ultralow voltages (e.g., 250 mV or lower) when compared to MOSFETs, and in particular, the operation in the MHz range is allowed even for voltage values in the order of 200–250 mV. This makes TFETs very well suited for applications where energy is crucial and low performance is tolerable, e.g., medical applications, sensor nodes, and implantable systems.

Another work [21] concluded that in SiGe-based TFET circuits, the minimum-energy point occurs at much lower bias in comparison to SOI/bulk MOSFETs (approximately 100 mV versus 250–350 mV). The minimum energy of TFETs is typically 35–85% better than conventional SOI/bulk MOSFETs mainly because of the significant reduction in the leakage energy contribution per cycle. The same work concludes that TFETs are less sensitive to variations in channel length and silicon thickness, which simplify the silicon printability issues arising at 32 nm and below.

In digital logic, the unidirectional conduction of TFET devices and the enhanced Miller capacitance of inverters can result in bootstrapped nodes within the circuit, causing potential failures and reliability risks if not properly handled by design techniques [1, 22]. In conventional MOSFETs, charges can be transferred under both positive and negative $V_{DS}$ bias due to a similar doping structure in the source and drain junctions. In contrast, a reverse-biased TFET has low conduction and cannot quickly dissipate the charge. This characteristic can result in switching nodes with transient "spikes," with voltage values above the power supply voltage and below ground.

With the unidirectional conduction of TFETs, modification of some circuits whose operation requires bidirectional conduction, such as conventional SRAM cells, is required. Proper SRAM cells designed with TFETs with a larger number of transistors can surpass the performance of conventional CMOS SRAM cells. In [10], the authors designed SRAM cells with heterojunction TFETs ($L_G$ = 32 nm), showing significant delay reduction below 0.4 V and dynamic energy reduction below 0.3 V due to the improved drive current of TFETs at low voltage when compared to Si-FinFET SRAM designs.

In Figure 4.10, the circuit schematic of an inverter driving another inverter is presented. In order to show the effect of the large $C_{GD}$ in TFETs (heterojunction TFET in this example) and consequent increase of Miller capacitance in TFET-based inverters, two squared input signals with two distinct frequencies are simulated with $f_1$ = 100 kHz (rise/fall time of 100 ns) and $f_2$ = 10 MHz (rise/fall time of 1 ns).

The p-type heterojunction TFETs ($L_G$ = 40 nm) are simulated with widths of 2 μm and the n-TFETs with widths of 1 μm. With a power supply voltage

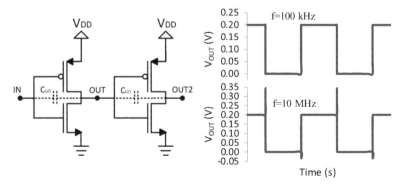

**Figure 4.10** Inverter circuit configuration designed with heterojunction tunnel field-effect transistors and transient output response.

of 0.2 V, one can see the effect of the Miller capacitance in the transient response of the first inverter, for an input signal with a frequency of 10 MHz. At lower frequencies and increased rise/fall time transitions, transient "spikes" during input signal transitions are not observed.

In order to evaluate the voltage range where TFETs are expected to present improved performance in digital cells when compared to conventional thermionic MOSFETs, one can compare the energy per clock transition of the inverter calculated as follows:

$$Energy\ per\ Clock\ Transition$$
$$= Dynamic\ Energy + Static\ Energy = \alpha C_L V_{DD}^2 + I_{lLEAK} V_{DD} \tau$$
(4.3)

In Equation (4.3), the energy per clock transition is characterized by a dynamic and static part. In the dynamic part, $\alpha$ is the activity factor (switching probability of the cell over a certain amount of time), $C_L$ the load capacitance at the output of the first inverter (capacitance at input of second inverter), and $V_{DD}$ the power supply voltage. In the static part, $I_{LEAK}$ is the static leakage current that equals the off-current of the transistor. The delay $\tau$ is calculated as the response time of the output (50% of its final value) when the input switches to 50% of its final value.

In Figures 4.11(a, b), the dynamic and static energies of the inverter configuration shown in Figure 4.10 are presented considering a switching activity factor of 0.01 and input signal frequency of 10 MHz. The results of the inverter circuit configuration designed with heterojunction and homojunction TFETs are normalized to those of FinFET-based inverters. It is shown that heterojunction and homojunction TFET-based inverters require less dynamic energy per clock transition than that of the FinFET counterpart at power supply voltages below 0.2 and 0.3 V, respectively. In Figure 4.11(c), it is shown that TFET-based inverters present a lower $C_L$ at low $V_{DD}$ in comparison to FinFET-based counterparts, thus requiring less dynamic energy consumption. In Figure 4.11(c), one can also observe that compared to the homojunction TFET-based inverter, the improved current capability of the heterojunction TFET-based counterpart at sub-0.15 V allows for lower delay times. This factor allied to the lower leakage current of heterojunction TFETs ($L_G$ = 40 nm) allows this technology to achieve a very low static energy when compared to the homojunction TFET counterpart and FinFET-based inverter as shown in Figure 4.11(b). As the leakage current of TFETs is shown much lower than that of conventional MOSFETs, the static energy of TFET-based

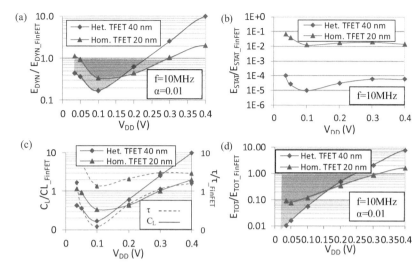

**Figure 4.11** Performance comparison of an inverter designed with heterojunction and homojunction tunnel field-effect transistors (normalized to the performance of FinFET-based inverter). (a) Dynamic, (b) static, (c) load capacitance and delay (50%), and (d) total energy per clock transition.

circuits can be reduced by several orders of magnitude, benefiting low power applications characterized by long waiting times, e.g., sensors triggered by sparse events. In Figure 4.11(d), the shaded regions show the voltage range where TFETs require less energy per clock transition in comparison to conventional thermionic MOSFETs.

## 4.4 TFETs in Analog Design

In this section, the heterojunction TFET ($L_G$ = 40 nm) and Si-FinFET ($L_G$ = 20 nm) are compared at a device level in order to analyze their impact on several FOM for analog design.

In Figure 4.12(a), the transconductance $g_m$ (channel width of 1 μm for TFET and 14 Fins for FinFET) is compared at different current levels. One can observe that at sub-10 μA, the TFET presents a larger transconductance than that of the FinFET. In conventional thermionic devices, the subthreshold region presents an SS constant over $V_{GS}$ (see Figure 4.4) and therefore, according to Equation (4.4), the transconductance of thermionic devices increases linearly with the required current [23]. In TFET devices, the SS

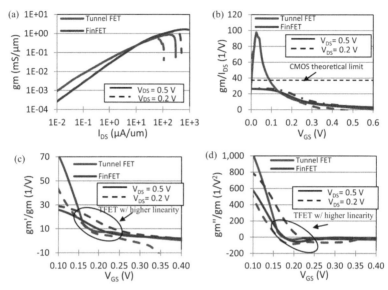

**Figure 4.12** Comparison of analog figures of merit between Si-FinFET and heterojunction tunnel field-effect transistor. (a) Transconductance. (b) Transconductance per current ratio. (c) Second and (d) third derivatives of current normalized to gm.

is shown dependent on $V_{GS}$, and therefore, the transconductance presents a nonlinear behavior in function of current.

$$gm_{subth} = \frac{ln\,(10)\,I_{DS}}{SS} \tag{4.4}$$

Considering the limited 60 mV/decade of SS at room temperature in conventional thermionic transistors, a $g_m/I_{DS}$ ratio of 38.5 $V^{-1}$ is settled as the theoretical limit for conventional thermionic technologies. Taking Equation (4.1) as reference, the transconductance efficiency $g_m/I_{DS}$ of TFET devices can be calculated as expressed by Equation (4.5) [24]. The transconductance efficiency of TFETs is therefore dependent on several factors such as function $f$ (capable of smoothly connecting the subthreshold and above-threshold operation regions), $V_{GS}$, electric field $F$, $b$ (material dependent), and tunneling window $\triangle\Phi$.

$$\frac{g_m}{I_{DS}} = \frac{1}{f}\frac{\partial f}{\partial V_{GS}} + \frac{1}{V_{TW}}\frac{\partial\triangle\Phi}{\partial V_{GS}} + \frac{1}{F}\frac{\partial F}{\partial V_{GS}} + \frac{b}{F^2}\frac{\partial F}{\partial V_{GS}} \tag{4.5}$$

As shown in Figure 4.12(b), TFET devices can achieve superior transconductance efficiency when compared to FinFETs at sub-0.1 V. This property is particularly interesting in low power design as circuits often require transistors to operate in the subthreshold region, i.e., where they are shown more efficient. In [14] and [15], the authors have shown that at low current bias, TFET-based amplifiers present a superior performance in comparison to conventional thermionic devices.

Despite the increased transconductance efficiency of TFETs at sub-0.1 V, the large $g'_m$ and $g''_m$ can degrade the linearity of circuits. A metric to determine the linearity of a device can be expressed by 4.6 [23]:

$$g'_m = \frac{\partial^2 I_{DS}}{\partial^2 V_{GS}} \quad \text{and} \quad g''_m = \frac{\partial^3 I_{DS}}{\partial^3 V_{GS}} \tag{4.6}$$

In Figures 4.12(c, d), one can observe the range of $V_{GS}$ (between 0.15 and 0.25 V) where the heterojunction TFET can enable circuits with larger linearity when compared to thermionic devices.

As mentioned in Section 4.2.2, the output characteristics of TFET devices present a large current saturation at large drain bias. This characteristic is observed when the allowed energy window in the source–channel interface reaches the maximum value set by $V_{GS}$. Consequently, the TFET output resistance is shown significantly larger than that of conventional thermionic devices.

In Figures 4.13(a, b), the output resistances of the heterojunction TFET and Si-FinFET are, respectively, presented. The large output resistance of TFET devices can improve the analog design at reduced technology nodes where the short-channel effects degrade the intrinsic gain of conventional technologies. In Figure 4.13(c), the heterojunction TFET device is shown to present a larger intrinsic gain compared to that of FinFET devices with a gate length of 20 nm. This characteristic is particularly interesting in ultralow power analog applications designed with short channel devices, in a way that the circuit complexity can be relaxed by the absence of cascaded stages required to increase the total circuit gain. Recently, a vertical nanowire heterojunction TFET designed with groups III–V materials (InAs–GaAsSb–GaSb) was shown to present an intrinsic gain of 2400 and transconductance efficiencies larger than 38 V$^{-1}$ (between 45 and 50 V$^{-1}$) [25]. These results are highly motivating for future ultralow power analog applications.

The combined effect of particular transconductance and intrinsic capacitances of TFETs can result in different cutoff frequencies $f_T$ when compared to those of FinFETs. The $f_T$ responses of the heterojunction TFET and Si-FinFET are calculated by Equation (4.7) and shown in Figure 4.14.

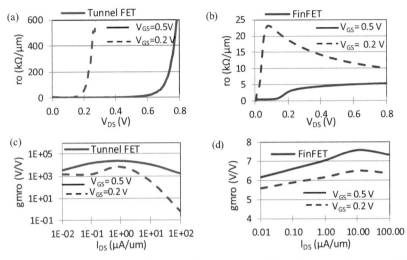

**Figure 4.13**   Output resistance of (a) heterojunction tunnel field-effect transistor (TFET) and (b) Si-FinFET. Intrinsic gain of (c) heterojunction TFET and (d) Si-FinFET.

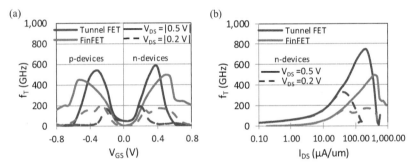

**Figure 4.14**   Unity gain frequency comparison of heterojunction tunnel field-effect transistor and Si-FinFET.

$$f_T = \frac{g_m}{2\pi\left(C_{GS} + C_{GD}\right)} \tag{4.7}$$

One can observe that the FinFET device can achieve an $f_T$ peak at approximately 500 GHz with large drain and gate bias. In contrast, the heterojunction TFET can achieve similar performances at lower gate bias and lower current. This characteristic can reduce the power consumption of TFET-based circuits by achieving a similar performance than conventional thermionic circuits at reduced power supply voltages.

## 4.5 TFETs' Circuit Layout Issues and Extra-parasitics

In TFET-circuit layout, one of the main difference in comparison to MOS-FETs is the nonidentical source and drain regions that require different doping types, doping levels, and materials [case of heterojunction TFETs, see Figure 4.15(a)]. As mentioned in [26], due to this characteristic, the TFET device can be fabricated using separate lithography steps for source and drain followed by etch and regrowth of the material. The TFET doping asymmetry also presents consequences in circuit layout density. In circuit applications, two MOSFET devices connected in series, i.e., drain of first device connected to the source terminal of second device can share a single contact as shown in Figure 4.15(b). In contrast, this layout is not possible in TFETs due to different materials used in the source and drain regions. As a major consequence, extra connections and footprint are required to achieve a series connection with TFETs. Vertical TFET structures as the one shown in Figure 4.15(a) are currently under investigation in order to reduce the device footprint area (and consequent circuit overhead compared to CMOS) and also due to the feasibility of the heterojunction structure implementation [9, 27]. Extra contacts in TFET-based circuits are expected to introduce parasitic capacitances that can jeopardize the performance of TFET-based circuits at ultralow power levels. In order to analyze such impact, improved models of TFETs and further investigation are required.

As an example, in [27], the authors have compared the layout and parasitic capacitances of heterojunction TFETs with FinFETs. They concluded that due to the vertical device structure of TFET, a smaller footprint can be observed in cells with small fan-in.

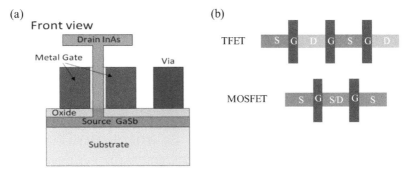

**Figure 4.15** (a) Vertical tunnel field-effect transistor (TFET) structure (adapted from [27]). (b) Additional contact region of TFETs due to nonsharing possibility of drain–source regions.

The area overhead can, however, lead to approximately 48% higher parasitic capacitances and resistances compared to FinFETs when the numbers of parallel and series connections increase. In their simulations, a 15-stage inverter-based ring oscillator shows a decreased performance of 8% when considering parasitics. Despite such degradation, the TFET-based circuit still presents less delay at power supply voltages below 0.45 V and large energy efficiency for power supply voltages in the range of 0.3–0.7 V.

## 4.6 Chapter Summary

In this chapter, the performance of a homojunction (InAs, $L_G$ = 20 nm) and heterojunction (InAs-GaSb, $L_G$ = 40 nm) TFET based on Verilog-A LUT models was analyzed and compared to the performance of a conventional Si-FinFET ($L_G$ = 20 nm) for digital and analog application purposes. The results show that the low leakage current and low intrinsic capacitance of TFETs can reduce the total energy consumption per clock transition (at sub-0.2 V) in digital cells when compared to the use of conventional thermionic technologies. It is also shown that due to the dominance of the gate-to-drain $C_{GD}$ capacitance in TFETs, the increased Miller capacitance of TFET-based inverter cells can lead to large "spikes" in the transient response of the inverter. This effect is more evident at low voltage operation (sub-0.2 V).

For analog applications, the low dependence of current on large drain bias results in TFETs with large output resistance and increase of intrinsic gain. This characteristic can benefit the design of analog applications with low technology nodes and reduce the complexity of circuits.

As the SS of TFETs changes according to $V_{GS}$, the transconductance efficiency of TFETs is not limited as in conventional thermionic devices. This behavior can enable TFETs with increased transconductance at lower current when compared to conventional MOSFETs.

As explained in Section 4.3, III–V TFETs still present a degraded SS in comparison to Si counterparts, and therefore, further investigation in III–V or novel materials that currently present high bulk and interface defects is required in order to achieve the performance shown by the simulated results of this chapter.

# References

[1] Lu, H., Ytterdal, T., and Seabaugh, A. "Universal TFET model." Available at: https://nanohub.org/publications/31, 2015.

[2] Lu, H., Kim, J. W., Esseni, D., and Seabaugh, A. "Continuous semiempirical model for the current-voltage characteristics of tunnel FETs," in *Proc. 15th Int. Conf. ULIS*, pp. 25–28, 2014.

[3] Lu, H., Esseni, D., and Seabaugh, A. "Universal analytic model for tunnel FET circuit simulation," *Solid State Electr.* 108, 110–117, 2015.

[4] Avci, U. E., Rios, R., Kuhn, K., and Young, I. A. "Comparison of performance, switching energy and process variations for the TFET and MOSFET in logic," in S*ymposium on VLSI Technology – Digest of Technical Papers*, pp. 124–125, 2011.

[5] Lu, Y., Zhou, G., Li, R., Liu, Q., Zhang, Q., Vasen, T., et al., "Performance of AlGaSb/InAs TFETs with gate electric field and tunneling direction aligned," in *IEEE Electron Device Letters*, vol. 33, no. 5, pp. 655–657, 2012.

[6] Liu, H., Saripalli, V., Narayanan, V., and Datta, S., "III–V tunnel FET model." Available at: https://nanohub.org/publications/12/2, 2015.

[7] Mookerjea, S., Krishnan, R., Datta, S., and Narayanan, V. "On enhanced Miller capacitance effect in interband tunnel transistors," in *IEEE Electron Device Letters*, vol. 30, no.10, pp. 1102–1104, 2009.

[8] Mookerjea, S., Krishnan, R., Datta, S., and Narayanan, V. "Effective capacitance and drive current for tunnel-FET (TFET) CV/I estimation," in *IEEE Transactions on Electron Devices*, vol. 56, no. 9, pp. 2092–2098, 2009.

[9] Datta, S., Liu, H., and Narayana, V. "Tunnel FET technology: A reliability perspective," *Microelectron. Reliabil.* 54, 861–874, 2014.

[10] Saripalli, V., Datta, S., Narayanan, V., and Kulkarni, J. "Variation-tolerant ultra low power heterojunction tunnel FET SRAM design," in *IEEE/ACM International Symposium on Nanoscale Architectures (NANOARCH)*, pp. 45–52, 2011.

[11] Cotter, M., Liu, H., Datta, S., and Narayanan, V. "Evaluation of tunnel FET-based flip-flop designs for low power, high performance applications," in *International Symposium on Quality Electronic Design (ISQED)*, pp. 430–437, 2013.

[12] Saripalli, V., Sun, G., Mishra, A., Xie, Y., Datta, S., and Narayanan, V. "Exploiting heterogeneity for energy efficiency in chip multiprocessors," in *Emerging and Selected Topics in Circuits and Systems, IEEE J.*, vol. 1, no. 2, pp. 109–119, 2011.

[13] Liu, H., Vaddi, R., Datta, S., and Narayanan, V. "Tunnel FET-based ultra-low power, high sensitivity UHF RFID rectifier," in *IEEE International Symposium on Low Power Electronics and Design (ISLPED)*, pp. 157–162, 2013.

[14] Trivedi, A., Carlo, S., and Mukhopadhyay, S. "Exploring tunnel-FET for ultra low power analog applications: a case study on operational transconductance amplifier," in *Design Automation Conference (DAC), ACM/EDAC/IEEE*, pp. 1–6, 2013.

[15] Liu, H., Datta, S., Shoaran, M., Schmid, A., Li, X., and Narayanan, V. "Tunnel FET-based ultra-low power, low-noise amplifier design for bio-signal acquisition," in *Low Power Electronics and Design, Int. Symp.*, 2014.

[16] Sinha, S., Yeric, G., Chandra, V., Cline, B., and Yu, C. "Exploring sub-20nm FinFET design with Predictive Technology Models," in *Design Automation Conference (DAC), 2012 49th ACM/EDAC/IEEE*, pp. 283–288, 2012.

[17] Sze, S. M., and Ng, K. K. Physics of Semiconductor Devices, 3rd Edn. NY: Wiley-Interscience, 2007.

[18] Seabaugh, A. C., and Zhang, Q. "Low-voltage tunnel transistors for beyond CMOS logic," in *Proceedings of the IEEE*, vol. 98, no. 12, pp. 2095–2110, 2010.

[19] Boucart, K., and Ionescu, A. "A new definition of threshold voltage in Tunnel FETs," *Solid-State Elec.* 52, 1318–1323, 2008.

[20] Alioto, M., and Esseni, D. "Performance and impact of process variations in tunnel-FET ultra-low voltage digital circuits," in *Proc. on 27th Symp. on Integrated Circuits and Syst. Design*, no. 32, 2014.

[21] Alioto, M., and Esseni, D. "Tunnel FETs for ultra-low voltage digital VLSI circuits: Part II–evaluation at circuit level and design perspectives," in *IEEE Transactions on Very Large Scale Integration (VLSI) Systems*, vol. 22, no. 12, pp. 2499–2512, 2014.

[22] Morris, D. H., Avci, U. E., Rios, R., and Young, I. A. "Design of low voltage tunneling-FET logic circuits considering asymmetric conduction characteristics," in *IEEE Journal on Emerging and Selected Topics in Circuits and Systems*, vol. 4, no. 4, pp. 380–388, 2014.

[23] Sedighi, B., Hu, X. S., Huichu, L., Nahas, J. J., and Niemier, M. "Analog circuit design using tunnel-FETs," in *Circuits and Systems I: Regular Papers, IEEE Trans.*, vol. 62, no. 1, pp. 39–48, 2015.

[24] Barboni, L., Siniscalchi, M., and Sensale-Rodriguez, B. "TFET-based circuit design using the transconductance generation efficiency gm/Id method," *IEEE Journal of the Electron Devices Society,* vol. 3, no. 3, pp. 208, 216, 2015.

[25] Memisevic, E., Svensson, J., Hellenbrand, M., Lind, E., Wernersson, L.-E. "Vertical InAs/GaAsSb/GaSb tunneling field-effect transistor on Si with S = 48 mV/decade and Ion = 10 $\mu$A/$\mu$m for Ioff = 1 nA/$\mu$m at Vds = 0.3 V," *IEEE International Electron Devices Meeting (IEDM),* San Francisco, CA, pp. 19.1.1–19.1.4, 2016.

[26] Avci, U. E., Morris, D. H., and Young, I. A. "Tunnel field-effect transistors: Prospects and challenges," in *IEEE Journal of the Electron Devices Society*, vol. 3, no. 3, pp. 88–95, 2015.

[27] Kim, M., et al. "Comparative Area and Parasitics Analysis in FinFET and Heterojunction Vertical TFET Standard Cells," *ACM J Emerg. Technol. Comput. Syst.* 12:38, 2016.

# 5

## Tunnel FET-based Charge Pumps

In this chapter, the performance of tunnel field-effect transistor (TFET)-based charge pumps suitable for energy harvesting (EH) applications is analyzed and compared to the performance of FinFET-based charge pumps at similar bias conditions. It is shown that due to the particular electrical characteristics of TFETs under reverse bias, the performance of conventional charge-pump topologies designed with TFETs degrades with the increase of power supply voltage and the decrease of output current. At a circuit level perspective, a possible solution to attenuate the reverse losses in TFET-based converters is the change of the gate bias magnitude in the reverse-biased TFETs. Therefore, in order to improve the efficiency of TFET-based charge pumps at a wider range of voltage and loads, different circuit topologies for the particular characteristics of this technology are proposed. All the simulated results are based on the Verilog-A-based lookup table (LUT) models of TFETs and predictive technology model of FinFETs described in Chapter 4.

### 5.1 Motivation

The interest in power supply circuits able to harvest energy from the surrounding environment for powering portable and wearable low-power systems has been increasing over the last years [1–4]. EH sources such as micro-photovoltaic cells [1, 2] and thermo-electric generators [3, 4] are characterized by extremely low output voltage and power values, thus preventing their direct usage in electronic applications. For this reason, and due to its easy implementation, charge pumps (also called switched-capacitor converters) have been widely considered to boost the output voltage of EH sources in order to meet the minimum supply voltage requirements of electronic systems.

Besides the inherent switching losses that characterize switching-based DC–DC converters, the main difficulty in achieving a good power conversion performance at low voltage (sub-0.25 V) and low power (submicrowatts) operation is related to the large conduction losses of conventional thermionic transistors applied in the conversion process and the reverse losses when the transistors operate during their off-state [5, 6]. Thermionic transistors such as FinFETs are characterized by a minimum SS of 60 mV/decade at room temperature. This characteristic limits the required current at low voltage operation in passive DC–DC converters, thus leading to increased forward losses in the transistors operating during on-states.

As explained in Chapter 4, TFET devices (in particular the heterojunction configuration) present improved electrical characteristics in comparison to conventional thermionic technologies at sub-0.25 V operation. On the other hand, TFETs are shown to conduct less current at larger values, and thereby their use is envisioned for low-voltage, low-performance applications. For this reason, under extremely low-voltage scenarios, TFETs appear as interesting devices to be implemented also in power conversion circuits. As an example, the work reported in [7] shows improvements in the performance of TFET-based charge-pump converters when compared to the application of the FinFET technology at sub-0.4 V levels. Such results are highly motivating for several applications where batteries are currently mandatory, thus benefiting from the low-voltage EH if the voltage requirements can be lowered.

Despite the improvements shown in the low-voltage conversion performance of the referenced work, the operation voltage range of the DC–DC converter is limited by the particular TFET electrical characteristics when the device is reverse biased. Under reverse-bias conditions, the intrinsic p-i-n diode of the TFET is forward biased and the reverse current is characterized either by reverse BTBT current at low reverse bias (occurring at the channel-drain interface) or drift–diffusion (DD) current at large reverse bias. While the second current mechanism is inevitable, the first can be attenuated by changing the magnitude of the gate bias when the TFET device is moderately reverse biased. This behavior was observed in several experiments with groups III–V TFETs [8–10].

For this reason, innovative DC–DC circuit topologies are required for TFETs in order to reduce the unwanted reverse current and therefore extend the voltage/power range operation of TFET-based converter circuits.

## 5.2 Problems Associated with TFETs in Charge Pumps

In order to understand the limitations of TFETs in conventional charge-pump topologies, one can consider as an example the gate cross-coupled charge-pump (GCCCP) topology shown in Figure 5.1(a). This charge pump is shown in the literature to produce the best performance at low-voltage operation in comparison to other charge-pump topologies [5], and therefore, it is denominated here as the conventional charge-pump topology. The principle of operation of the GCCCP converter can be divided into two regions of operation as shown in Figure 5.1(c). In region I, the low to high transition of Clock 1 increases the voltage of node *Int1* to $2V_{DD}-V_{DS1}$. During the same time, the voltage at node *Int2* is reduced to $V_{DD}-V_{DS2}$. In this region, transistors M1 and M4 are reverse biased, i.e., operate in an off-state while transistors M2 and M3 are forward biased (on-state). Considering steady-state conditions, Table 5.1 presents the bias characteristics of TFETs in the first region of operation.

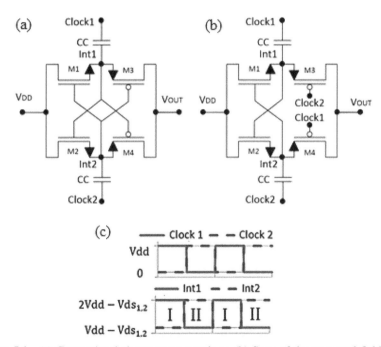

**Figure 5.1** (a) Conventional charge-pump topology. (b) State-of-the-art tunnel field-effect transistor-based charge pump [7]. (c) Regions of operation.

**Table 5.1**   Bias conditions of the TFETs applied in the GCCCP considering Region I

| Reg. I | State | $V_{GS}$ | $V_{DS}$ |
|---|---|---|---|
| M1 (n) | Off | Int2 – Int1 = $-V_{DD}$ | $V_{DD}$ – Int1 = $-V_{DD} + V_{DS1}$ |
| M2 (n) | On | Int1 – Int2 = $V_{DD}$ | $V_{DD}$ – Int2 = $V_{DS2}$ |
| M3 (p) | On | Int2 – Int1 = $-V_{DD}$ | $V_{OUT}$ – Int1 = $-V_{SD3}$ |
| M4 (p) | Off | Int1 – Int2 = $V_{DD}$ | $V_{OUT}$ – Int2 = $V_{DD} - V_{SD3}$ |

**Table 5.2**   Bias conditions of the TFETs applied in the GCCCP considering Region II

| Reg. II | State | $V_{GS}$ | $V_{DS}$ |
|---|---|---|---|
| M1 (n) | On | Int2 – Int1 = $V_{DD}$ | $V_{DD}$ – Int1 = $V_{DS1}$ |
| M2 (n) | Off | Int1 – Int2 = $-V_{DD}$ | $V_{DD}$ – Int2 = $-V_{DD} + V_{DS2}$ |
| M3 (p) | Off | Int2 – Int1 = $V_{DD}$ | $V_{OUT}$ – Int1 = $V_{DD} - V_{SD4}$ |
| M4 (p) | On | Int1 – Int2 = $-V_{DD}$ | $V_{OUT}$ – Int2 = $-V_{SD3}$ |

In the second region of operation, the high to low (low to high) transition of Clock 1 (Clock 2) and consequent reduction (increase) of voltage in node *int1* (*int2*) result in a forward bias condition of transistors M1 and M4 and reverse bias in transistors M2 and M3. In Table 5.2, the bias characteristics of TFETs operating in the CGCCP during the second region of operation are presented.

As previously explained in Section 4.2.2, under reverse-bias conditions (negative $V_{DS}$ for n-type and positive $V_{DS}$ for p-type TFET), the intrinsic p-i-n diode of the TFET is forward biased and the resulting reverse current is characterized by two different carrier injection mechanisms. In n-TFETs, the reverse current at low reverse-bias condition ($V_{DS} < 0$ V) is due to a reverse BTBT carrier mechanism occurring at the channel–drain interface as shown in Figure 5.2(a). With the increase of reverse bias ($V_{DS} \ll 0$ V), the BTBT mechanism is suppressed due to the increase of energy bands in the drain region, and the reverse current is characterized by excess and drift–diffusion as shown in Figure 5.2(b). The same behavior characterizes p-TFETs with positive $V_{DS}$.

As the transistors applied in charge pumps operate at forward (on-state) and reverse (off-state) bias during consecutive time intervals, it is of the major importance to mitigate the reverse current conducted by TFETs during their off-state operation.

According to Tables 5.1 and 5.2 and considering no forward losses in the transistors, the reverse bias conditions of the TFET transistors are always characterized by a $V_{GS} = V_{DS}$ condition. In Figure 5.3, the reverse current of reverse-biased heterojunction (a) and homojunction (b) n-TFETs with this bias condition is shown. One can observe that the reverse current of TFETs

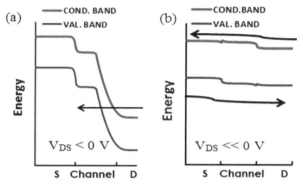

**Figure 5.2** Tunnel field-effect transistor (TFET) energy band diagram of an n-TFET under reverse-bias conditions. (a) Reverse BTBT mechanism. (b) Drift–diffusion mechanism.

increases with the magnitude of the reverse drain bias, and therefore, their use in charge pumps is limited only to low-voltage operation. As a basis of comparison, the reverse current of a Si-FinFET ($L_G$ = 20 nm) applied in the GCCCP topology is compared to both TFET structures.

The shadow regions shown in Figure 5.3 show the voltage range where TFETs applied in the GCCCP present reduced reverse current (and consequent reduction of reverse losses) compared to the use of thermionic FinFETs. As shown in Figure 5.3(a), the reverse current of a reverse-biased heterojunction TFET (HTFET) can be controlled by the gate bias magnitude [8–10]. It is shown that with a $V_{GS}$ = 0 V, the reverse current is attenuated for a significant

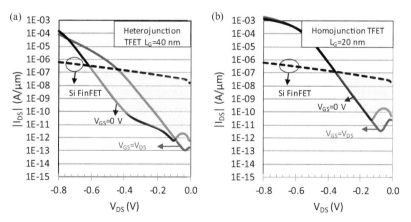

**Figure 5.3** Reverse current comparison of (a) heterojunction n-tunnel field-effect transistor (TFET) and (b) homojunction n-TFET during the reverse bias state.

range of $V_{DS}$ (−0.1 to −0.6 V). With a similar source–channel region, this characteristic is not observed in the homojunction TFET counterpart. For this reason, as the magnitude of reverse current with HTFETs can be controlled by the magnitude of the gate bias, changes in the conventional charge-pump topology are required in order to reduce the reverse losses of the converter and improve the conversion efficiency at a wider range of voltage operation.

In [7], the authors presented changes in the conventional GCCCP, in order to alleviate the reverse losses produced by the reverse-biased HTFETs. In their charge-pump topology, the gate control signals of the two p-type transistors are redirected to the bottom of the two coupling capacitors as shown in Figure 5.2(c). This solution forces the $V_{GS}$ of HTFETs M3 and M4 to, respectively, $V_{DS1}$ and $V_{DS2}$ (approximately 0 V when the required output current is low) when the transistors are reverse biased. This solution also lowers the conduction losses of transistors M3 and M4 by applying a larger $V_{GS}$ magnitude when these transistors are forward biased. This topology solution presents, however, some limitations: larger output currents produce larger conduction losses in the input transistors M1 and M2 (larger $V_{DS1}$ and $V_{DS2}$) and this way, the magnitude of $V_{GS}$ in reverse-biased TFETs will deviate from 0 V, and therefore, a different solution is required. Also, the reverse current of M1 and M2 (under reverse bias conditions) is not solved. In the following subsection, circuit solutions to attenuate the reverse current of HTFETs in charge pumps are presented.

## 5.3 Circuit-level Solutions for Reverse-biased TFETs

At a circuit level, a possible solution to attenuate the reverse current of HTFETs operating under reverse-biased conditions (off-state) is to set their $V_{GS}$ magnitude to 0 V. To perform this behavior, auxiliary transistors and capacitors can be used as shown in Figure 5.4. Considering as an example the n-TFET M1, the auxiliary transistor (*Maux*) and capacitor (*Caux*) are required to fix the gate node of M1with a voltage value equal to the highest voltage value of nodes *Int1* and *Int2*. For a single-stage converter, this value is ideally equal to twice the voltage of the stage input, and for multiple stages, the value equals the output voltage of the respective stage.

As shown by the transient behavior of Figure 5.4, the proposed solution applies a $V_{GS}$ with a magnitude close to 0 V when the transistor M1 is reverse biased (neglecting the forward losses of the auxiliary transistor)

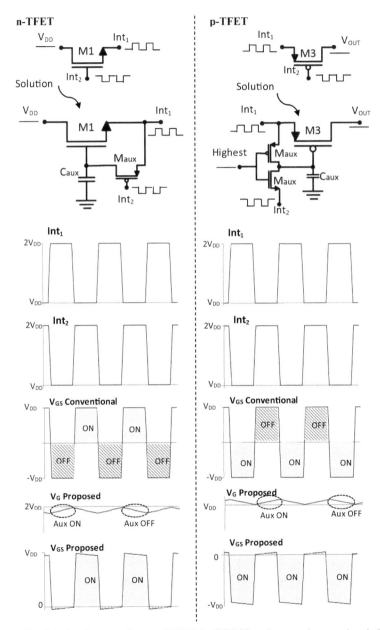

**Figure 5.4** Solution for n- and p-type TFETs in GCCCP and respective transient behavior.

and a positive $V_{GS}$ magnitude (ideally the same of the conventional GCCCP) when the transistor is forward based. A similar solution is proposed for p-TFET devices. In this case, an auxiliary inverter is required to fix the gate voltage of p-device M3 with a value equal to the lowest voltage value of nodes *int1* and *int2*. The input of the inverter is biased with a voltage equal to the highest voltage of nodes *int1* and *int2*. As shown by the transient behavior of Figure 5.4, the proposed solution applies a $V_{GS}$ magnitude close to 0 V when the p-device is reverse biased, and negative $V_{GS}$ (ideally the same of the conventional GCCCP) when the p-device is forward biased.

## 5.4 Proposed TFET-based Charge Pump

In Figure 5.5, a TFET-based charge pump is proposed. The gates of the main TFET transistors M1 and M2 are connected and biased with the auxiliary transistor *M1aux* and capacitor *C1aux*, while the gates of the transistors M3 and M4 are biased by an auxiliary inverter (*M2aux* and *M3aux*) and capacitor *C2aux*. During the first region of operation (*Vint1* > *Vint2*), both the auxiliary transistor *M1aux* and the auxiliary inverter are active, thus charging the capacitors at nodes *int2** and *int1** to the voltage values of respective nodes *int1* ($\approx 2V_{DD}$) and *int2* ($\approx V_{DD}$) (neglecting the forward losses of the main and auxiliary transistors). During this region, the reverse-biased transistors M1 and M4 present a $V_{GS} \approx 0$ V and the forward-biased transistors M2 and M3 present, respectively, $V_{GS} \approx V_{DD}$ and $V_{GS} \approx -V_{DD}$.

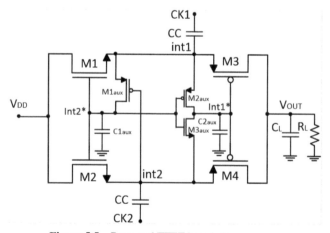

**Figure 5.5**    Proposed TFET-based charge pump.

During the second region of operation, the auxiliary transistors are reverse biased (off-state) and the voltage values at nodes *int1\** and *int2\** are retained. Now transistors M2 and M3 are reverse biased with a $V_{GS} \approx 0$ V and transistors M1 and M4 are forward biased with, respectively, $V_{GS} \approx V_{DD}$ and $V_{GS} \approx -V_{DD}$. The $V_{DS}$ values applied to each main transistor during each region of operation remain the same as those shown in Tables 5.1 and 5.2. The $V_{GS}$ values applied to the main transistors are presented in Table 5.3. In order to understand the operation of the proposed TFET-based charge-pump operation, Figure 5.6 shows the transient behavior of the internal nodes inside the charge-pump stage, considering as an example an operation frequency of 100 MHz and power supply voltage of 160 mV.

**Table 5.3** $V_{GS}$ bias of the proposed TFET-based charge pump

|          | $V_{GS}$ Region I              | $V_{GS}$ Region II              |
|----------|-------------------------------|--------------------------------|
| M1 (n)   | Int2\* – Int1 $\approx 0$      | Int2\* – Int1 $\approx V_{DD}$   |
| M2 (n)   | Int2\* – Int2 $\approx V_{DD}$ | Int2\* – Int2 $\approx 0$         |
| M3 (p)   | Int1\* – Int1 $\approx -V_{DD}$ | Int1\* – Int1 $\approx 0$        |
| M4 (p)   | Int1\* – Int2 $\approx 0$       | Int1\* – Int2 $\approx -V_{DD}$  |

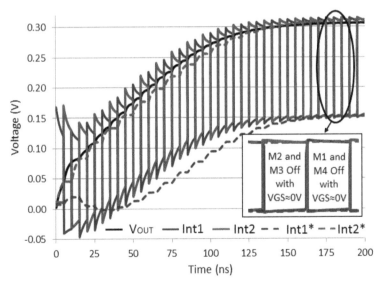

**Figure 5.6** Transient behavior of the proposed tunnel field-effect transistor (TFET)-based CP with heterojunction III–V TFETs considering $I_{OUT} = 1$ μA, WM1-4 = 1 μm, Waux = 100 nm, $f$ = 100 MHz, $C_L$ = CC = 1 pF, Caux = 0.1 pF, and $V_{DD}$ = 160 mV.

## 5.5 Capacitance Optimization of Charge-pump Stage

In this section, the load $C_L$ and coupling capacitances CC of the proposed charge pump (Figure 5.5) designed with HTFETs are optimized for increased power conversion efficiency (PCE). The optimized capacitance values are then applied to the conventional and state-of-the-art (SOA) HTFET-based charge pump shown, respectively, in Figures 5.1(a, b). In order to evaluate the performance of the charge-pump for low and high frequencies of operation, two distinct clock frequencies are chosen: 1 KHz and 100 MHz. For simplification, the transition times of clocks (high-to-low and low-to-high) are simulated as 1% of the clock period. The transient behavior of the clocks is presented in Figure 5.7 (not at scale). As the objective is to estimate the performance of the charge-pump stage and not the clock circuitry, ideal clock phases are simulated by two ideal voltage sources.

In the simulations, the PCE of the charge-pump stage is calculated as expressed by (5.1). In order to increase the PCE of the charge-pump stage, the power losses (*Plosses*) have to be minimized. According to (5.2), the reverse, conduction, and switching losses are presented as the major source of losses that degrade the conversion efficiency of a converter stage.

$$PCE = \frac{P_{IN,\,DC} - P_{losses}}{P_{IN,\,DC}} = \frac{P_{OUT,\,DC}}{P_{IN,\,DC}} = \frac{V_{OUT_{DC}} I_{OUT_{DC}}}{V_{IN_{DC}} I_{IN_{DC}}} \quad (5.1)$$

$$P_{losses} = P_{reverse} + P_{conduction} + P_{switching} \quad (5.2)$$

The first power loss is characterized by the current that flows from the output to the input of the stage due to the non-fully-closed transistors. With the different electrical characteristics of TFETs under reverse bias conditions, this source of losses is important at large voltage values. Conduction losses

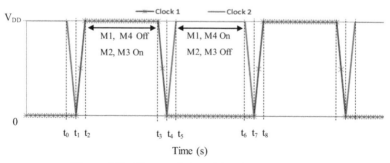

**Figure 5.7**   Characteristics of Clock 1 and Clock 2.

exist due to a non-zero channel resistance in the transistors biased under forward bias conditions. For a specific channel width, the increase of current conduction results in increased forward voltage drop and a consequent increase of conduction losses. The increase of the transistor width is presented as a possible solution to attenuate the conduction losses; however, larger transistor sizes result in larger parasitic capacitances, switching, and reverse losses.

The PCE of the proposed charge pump designed with HTFETs (HTFET-Prop. CP) is simulated for the two frequencies in study, considering an ideal input voltage source $V_{DD}$ of 0.4 V and an output load of 100 k$\Omega$. The values of coupling capacitances CC are equal to the values of the load capacitance $C_L$. The main TFET devices M1–M4 are simulated with $L_G$ = 40 nm and widths of 1 $\mu$m. The auxiliary transistors are simulated with widths of 100 nm (1/10 the size of the main transistors) and the auxiliary capacitors with capacitance values equal to 1/10 of the CC values.

In Figure 5.8, one can observe that there is a range of capacitance values that maximizes the PCE of the charge-pump stage. For both frequencies of operation, low CC values result in low pumped charge capability of the coupling capacitors to drive the load. The consequent pumping inefficiency results in degraded voltage across the load. According to the results and for the remaining simulations, load capacitors $C_L$ with 100 nF and 10 pF are chosen for charge-pump frequencies of 1 kHz and 100 MHz, respectively.

In Figure 5.9, the PCE of the proposed charge pump is presented as a function of the CC capacitance (considering $C_L$ values previously defined). For the two clock frequencies in study, two distinct loads (100 k$\Omega$ and 1 M$\Omega$) and input voltages (0.2 and 0.4 V) are considered. As previously explained,

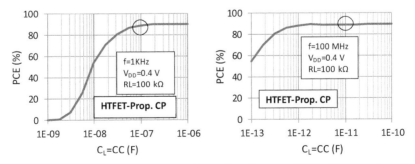

**Figure 5.8** Power conversion efficiency of heterojunction tunnel field-effect transistor-Prop. Cp with one-stage as a function of load capacitor.

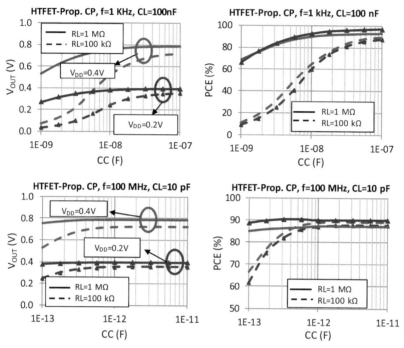

**Figure 5.9**    Power conversion efficiency of heterojunction tunnel field-effect transistor-Prop. CP with one-stage as a function of coupling capacitors.

increasing the CC capacitance allows for increased voltage values across the load ($V_{OUT}$) and PCE. Coupling capacitances of 100 nF and 1 pF for clock frequencies of, respectively, 1 kHz and 100 MHz are shown to produce large PCE values, and therefore these values are used as reference in the following section.

## 5.6 Charge Pumps' Performance Comparison

In this section, the performance of the conventional [Figure 5.1(a)], SOA [Figure 5.1(b)], and proposed charge pump (Figure 5.5) is compared. The three topologies are simulated with HTFETs ($L_G$ = 40 nm), and for a basis of comparison, simulations of the conventional charge-pump topology designed with thermionic transistors (FinFET $L_G$ = 20 nm) are included. To highlight the advantage of using TFET devices for EH applications, the electrical characteristics of a commercial ultralow thin-film thermo-generator, in particular the MPG-D655 from Micropelt [11], are used to simulate the input power

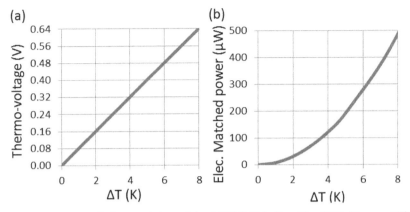

**Figure 5.10**   Electrical characteristics of the MPG-D655, $U$ = 80 mV/K ($T_{amb}$ = 25°C). RTH = 22 K/W and Relec = 210 $\Omega$.

supply voltage of the charge pumps. The thermo-generator presents a Seebeck coefficient of 80 mV/K and an electrical resistance of 210 $\Omega$. The electrical characteristics of the thermo-generator are shown in Figure 5.10.

Figures 5.11 and 5.12 show a comparison between the performances of the proposed (HTFET Prop. CP), conventional (HTFET GCCCP), and SOA (HTFET SOA CP) charge-pump topologies designed with HTFETs for the two clock frequencies under study: 1 kHz and 100 MHz. For comparison purposes, the performance of the conventional charge pump designed with FinFETs (Nfins = 14) is included. Two different variations of temperature in the thermo-generator are considered: $\Delta K$ = 2 K ($V_{DD}$ = 160 mV) and $\Delta K$ = 6 K ($V_{DD}$ = 480 mV).

One can observe that the performance of the FinFET-based charge pump is worse compared to the TFET-based counterparts (at both frequencies of operation) when the power supply voltage $V_{DD}$ and required output currents are low (due to larger reverse losses of FinFETs in the voltage range considered, see Figure 5.3). In contrast, the performance of the FinFET-based charge pump is better at large power supply voltage ($V_{DD}$ = 480 mV or above) and at large required output current. This is directly related to the improved performance of conventional thermionic devices at large voltage (improved drive current), when compared to TFETs.

On the other hand, the performance of TFET-based charge pumps is better at low-voltage/low-current operation. For both frequencies of operation, the HTFET-GCCCP converter presents the largest PCE values at submicrowatt operation at $V_{DD}$ = 160 mV. In contrast, the proposed TFET-based charge

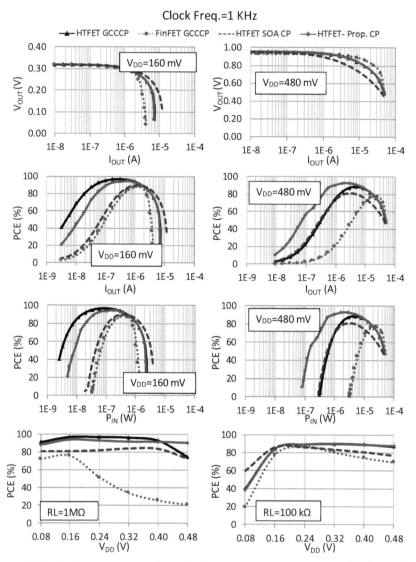

**Figure 5.11** Performance comparison of charge pumps (one stage) considering a clock frequency of 1 KHz. CC = $C_L$ = 100 nF. TFET-based charge pumps: WM1–M4 = 1 $\mu$m, *Waux* = 0.1 $\mu$m, and *Caux* = 10 nF. FinFET-based charge pump: Nfins M1–M4 = 14.

pump shows improved efficiency at larger power supply voltage ($V_{DD}$ = 480 mV) and sub-10 $\mu$W operation. This is directly related to the reduction of reverse loss characteristic of the proposed charge-pump topology

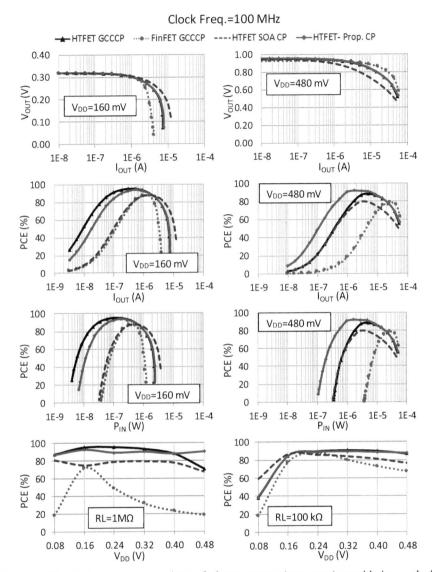

**Figure 5.12** Performance comparison of charge pumps (one stage) considering a clock frequency of 100 MHz. CC = 1 pF and $C_L$ = 10 pF. Tunnel field-effect transistor-based charge pumps: WM1–M4 = 1 μm, $W_{aux}$ = 0.1 μm, and $C_{aux}$ = 0.1 pF. FinFET-based charge pump: Nfins M1–M4 = 14.

when the TFETs are under high reverse bias state. The small degradation of the proposed charge-pump topology at low voltage (sub-0.4 V) and low current (submicroamperes) is directly related with the switching losses produced by the auxiliary circuitry during the clock transitions, i.e., during *t0–t2* and *t3–t5* (see Figure 5.7).

At very low voltage (sub-160 mV) and for large output currents, the HTFET SOA CP produces the largest output voltage values and power efficiencies due to the reduction of the conduction losses in the output transistors M3 and M4 when subjected to forward bias conditions. However, when the required output current is low, the switching losses produced by the output transistors (larger than the TFET-based CP counterparts due to larger gate voltage magnitudes) in this charge-pump topology degrade the PCE of the stage.

In Figure 5.13, the distribution of power losses in each charge-pump stage is presented for a common load current of 1 μA and the two clock frequencies under study. One can observe that for a small temperature difference between

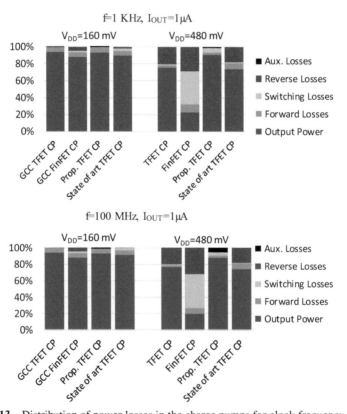

**Figure 5.13**   Distribution of power losses in the charge pumps for clock frequency of 1 kHz and 100 MHz and load of 1 μA.

**Table 5.4** Suitable charge-pump topology for different voltage/power ranges

|  | Ultra-Low Voltage | | Low-Voltage | | Medium Voltage | |
|---|---|---|---|---|---|---|
| Voltage Range | <160 mV | | 160–480 mV | | >400 mV | |
| $P_{IN}$ | <1μW | >1μW | <1μW | >1μW | <10μW | >10μW |
| Suitable Topology | HTFET-GCCCP | HTFET-SOA CP | HTFET-GCCCP | HTFET-GCCCP HTFET-Prop. CP | HTFET. Prop. CP | FinFET-GCCCP |

the plates of the thermo-generator source ($\Delta K = 2$ K, $V_{DD} = 160$ mV) and for both clock frequencies, the conventional and proposed charge pumps enable the largest power to the load. At a larger variation of temperature ($\Delta K = 6$ K, $V_{DD} = 480$ mV), one can observe that for both frequencies of operation, the proposed charge pump presents the largest power to the load (output power). When compared to conventional and SOA HTFET-based charge pumps, the reverse losses are highly reduced. At a frequency of 100 MHz, the losses caused by the auxiliary circuitry are shown to increase, thus degrading the PCE of the converter when compared to a lower frequency of operation.

Table 5.4 summarizes the most suitable (larger efficiency) charge-pump topology for different power supply voltage values. The proposed HTFET-based charge pump is shown as a suitable topology for input voltages larger than 400 mV and sub-10 μW power operation.

## 5.7 Chapter Summary

In this chapter, a charge pump designed for optimized operation with TFETs is presented. It is shown that when the HTFET is reverse biased, the reverse current magnitude can be attenuated by changing the $V_{GS}$ magnitude to 0 V. As the conventional GCCCP topology characterizes reverse-biased TFETs with a $|V_{GS}| = |V_{DS}|$, a different topology is required. The proposed charge pump includes an auxiliary circuitry that applies a $V_{GS} = 0$ V when the n-type and p-type HTFETs are reverse biased, and $V_{GS}$ magnitudes equal to those applied in the GCCCP topology when the devices are forward biased.

In order to minimize the increased layout footprint of the proposed charge pump, the size ratio of the auxiliary transistors and capacitors is chosen as 1/10 the size of the main transistors and coupling capacitors, respectively.

For comparison purposes, the results are presented considering single-stage charge pumps and unregulated output voltage. It is shown by simulations that power supply voltages above 0.4 V characterize the proposed charge-pump topology with improved power conversion efficiencies in comparison to the counterpart topologies. Despite larger switching losses caused by the auxiliary circuit, the improved efficiency of the proposed charge pump is due to the reduction of reverse losses suffered by the main transistors inside the stage when subjected to large reverse bias.

## References

[1] Brunelli, D., Moser, C., Thiele, L., and Benini, K., "Design of a solar-harvesting circuit for Batteryless embedded systems," in *IEEE Transactions on Circuits and Systems I: Regular Papers*, vol. 56, no. 11, pp. 2519–2528, 2009.

[2] Hsu, S. W., Fong, E., Jain, V., Kleeburg, T., and Amirtharajah, R. "Switched-capacitor boost converter design and modeling for indoor optical energy harvesting with integrated photodiodes," *Int. Symposium on Low Power Electronics and Design*, pp. 169–174, 2013.

[3] Mateu, L., Codrea, C., Lucas, N., Pollak, M. and Spies, P. "Human Body Energy Harvesting Thermogenerator for Sensing Applications," in *International Conference on Sensor Technologies and Applications,* pp. 366–372, 2007.

[4] Bassi, G., Colalongo, L., Richelli, A., and Kovács-Vajna, Z., "A 150mV-1.2V fully-integrated DC-DC converter for Thermal Energy Harvesting," in *International Symposium on Power Electronics Power Electronics, Electrical Drives, Automation and Motion*, pp. 331–334, 2012.

[5] Wong, O. Y., Wong, H., Tam, W. S., and Kok, C. W. "A comparative study of charge pumping circuits for flash memory applications," *Microelectron. Reliabil.* vol. 52, no. 4, pp. 670–687, 2012.

[6] Palumbo, G., and Pappalardo, D. "Charge pump circuits: An overview on design strategies and topologies," in *IEEE Circuits and Systems Magazine*, vol. 10, no. 1, pp. 31–45, 2010.

[7] Heo, U., Li, X., Liu, H., Gupta, S., Datta, S., and Narayanan, V. "A high-efficiency switched-capacitance HTFET charge pump for low-input-voltage applications," *VLSI Design (VLSID)*, pp. 304–309, 2015.

[8] Memisevic, E., Svensson, J., Hellenbrand, M., Lind, E., and Wernersson, L.-E. "Vertical InAs/GaAsSb/GaSb tunneling field-effect transistor on Si with S = 48 mV/decade and Ion = 10 μA/μm for Ioff = 1 nA/μm at Vds = 0.3 V," in *IEEE International Electron Devices Meeting (IEDM)*, San Francisco, CA, pp. 19.1.1–19.1.4, 2016.

[9] Pandey, R., Madan, H., Liu, H., Chobpattana, V., Barth, M., Rajamohanan, B., et al. "Demonstration of p-type In0.7Ga0.3As/GaAs0. 35Sb0.65 and n-type GaAs0.4Sb0.6/In0.65Ga0.35As complimentary Heterojunction Vertical Tunnel FETs for ultra-low power logic," in *Symposium on VLSI Technology (VLSI Technology)*, pp. T206–T207, 2015.

[10] Pandey, R., Schulte-Braucks, C., Sajjad, R. N., Barth, M., Ghosh, R. K., Grisafe, B., et al. "Performance benchmarking of p-type In0.65Ga0.35As/GaAs0.4Sb0.6 and Ge/Ge0.93Sn0.07 hetero-junction tunnel FETs," *IEEE International Electron Devices Meeting (IEDM)*, San Francisco, CA, pp. 19.6.1–19.6.4, 2016.

[11] "MPG-D655 Thin Film Thermogenerator Preliminary Datasheet," Accessed January 2014, Available at: http://micropelt.com/downloads/ datasheet_mpg_d655.pdf

# 6

## Tunnel FET-based Rectifiers

In this chapter, the performance of tunnel field-effect transistor (TFET)-based rectifiers is explored for ultra-low power applications and compared to conventional thermionic device-based rectifiers at similar bias conditions. In order to counteract the reverse current conducted by reverse-biased TFETs (intrinsic p-i-n diode is forward biased), different rectifier topologies are proposed and compared. All the simulated results are based on the Verilog-A based lookup table (LUT) TFET models and predictive technology model-based FinFET models described in Chapter 4.

## 6.1 Motivation

Several low-power applications can benefit from the surrounding radiated energy in order to power their circuits. Radio-frequency (RF) identification tags and biomedical implants are examples of RF-powered circuits that can be placed in areas of difficult access. As the constant replacement of their batteries is undesired, the field of energy harvesting from ambient has gained importance as shown by recent works [1–3].

One of the main limitations of RF-powered circuits is the low efficiency at low RF input power levels (submicrowatts). This is directly related to the low efficiency shown by the front-end rectifier at very low-voltage levels. As conventional rectifiers are designed with thermionic metal-oxide-semiconductor field-effect transistors (MOSFETs), the performance degrades with the decrease of induced RF voltage at the rectifier input terminals [4–8]. As TFETs present improved electrical performance at sub-0.25 V, it is of interest to study the performance of this technology in low-power/low-voltage rectifiers. In [9], the authors have shown by simulations that TFET-based passive rectifiers present improved rectification efficiency at sub-30 dBm in comparison to FinFET-based rectifiers.

Despite the advantages of using TFETs in rectifiers shown by the authors of the referenced work, the direct replacement of conventional thermionic transistors by the TFET technology is, in some cases, not appropriate. As explained in the previous chapter, under reverse bias conditions, TFETs present particular electrical characteristics. Reverse BTBT and drift–diffusion (DD) carrier injection mechanism at low and high reverse bias, respectively, can degrade the performance of rectifiers due to the consequent reverse losses. While the second carrier mechanism is inevitable, the reverse current due to the first mechanism can be attenuated by different rectifier topologies by changing the gate magnitude of reverse-biased TFETs. Therefore, in this chapter, an innovative TFET-based rectifier is proposed and explored at different induced RF voltage and power levels.

## 6.2 State-of-the-art TFET-based Rectifier

The gate cross-coupled rectifier (GCCR) presented in Figure 6.1(a) has been the topology of choice by several works due to its easy implementation and good results at low-voltage/power-applications [6–8]. In [9], the GCCR designed with heterojunction TFET (HTFET) devices (GaSb-InAs) was shown by the authors to present a better power conversion efficiency (PCE) at a wider range of voltage/power operation in comparison to other rectifier topologies (PCE >50% at RF input power between −40 and −25 dBm). Despite the good performance shown at low power operation, the voltage/power range of TFET-based rectifiers can be improved by reducing the reverse losses of individual devices inside the rectifier stage during their off-state conditions (reverse-biased state).

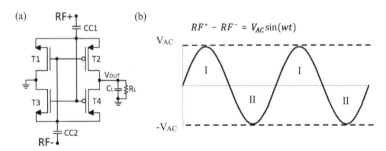

**Figure 6.1**   (a) Conventional gate cross-coupled rectifier and (b) its two different regions of operation.

Assuming that the RF signal presents a sinusoidal behavior such as the one shown in Figure 6.1(b), the GCCR operation can be divided into two regions: region I where the voltage at node $RF^+$ is larger than that of node $RF^-$ and region II where the opposite condition applies. In Figure 6.2, the transient behavior of nodes $RF^+$ and $RF^-$ is divided into different time intervals ($t0 \rightarrow t6$). During each time interval, the TFET devices in the GCCR are characterized as follows:

- $t0 \rightarrow t1$ and $t2 \rightarrow t3$: devices T2 and T3 are reverse biased with their $V_{GS}$ and $V_{DS}$ presenting opposite polarities (T2 and T3 OFF);
- $t1 \rightarrow t2$: time interval where the voltage at node $RF^+$ is larger than the output voltage of the rectifier stage by the threshold voltage of device T2 (T2 ON); during the same time interval, the voltage at node $RF^-$ is lower than the input voltage of the stage by the threshold voltage of T3 (T3 ON);
- $t0 \rightarrow t3$: during this time interval, devices T1 and T4 are reverse biased (OFF) with $V_{DS}$ and $V_{GS}$ presenting a similar polarity: "$-$" for n-type and "$+$" for p-types;
- $t4 \rightarrow t5$: time interval where the voltage at node $RF^+$ is lower than the input of the rectifier and voltage at node $RF^-$ is larger than the output of the rectifier by, respectively, the threshold voltage of T1 and T4 (T1 and T4 ON);

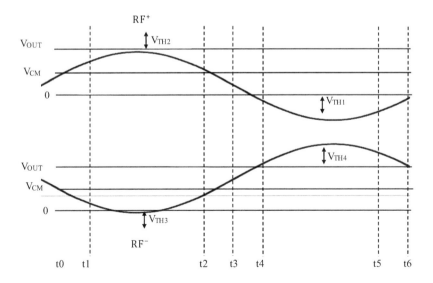

**Figure 6.2** Transient behavior of $RF^+$ and $RF^-$ nodes.

- $t3 \rightarrow t4$ and $t5 \rightarrow t6$: devices T1 and T4 are reverse biased with their $V_{GS}$ and $V_{DS}$ presenting similar polarities (T1 and T4 OFF);
- $t3 \rightarrow t6$: during this time interval, devices T2 and T3 are reverse biased (OFF) with $V_{DS}$ and $V_{GS}$ presenting similar polarities: "−" for n-type and "+" for p-types.

The bias conditions of a TFET-based GCCR according to the different time intervals described are summarized in Tables 6.1 and 6.2.

Similar to the charge-pump topology presented in the previous chapter, the reverse, conduction, and switching power are presented as the main losses in the rectification process. The PCE of the rectifier stage is calculated as shown by (6.1).

$$PCE = \frac{P_{OUT,DC}}{P_{IN,RF}} = \frac{V_{OUT,DC} \times I_{OUT,DC}}{\frac{1}{T} \int_0^T V_{AC}sin(wt)I_{AC}sin(wt)dt} \tag{6.1}$$

In Figure 6.3, the performance comparison of the GCCR designed with HTFETs (InAS-GaSb, $L_G$ = 40 nm) considering two frequencies of operation (100 and 915 MHz) and loads (RL = 100 k$\Omega$ and RL = 10 k$\Omega$) is presented.

**Table 6.1**    Steady-state bias conditions of the TFET-GCCR in region I

| Region I | $V_{GS}$ | $V_{DS}$ | | $\Delta T$ | State |
|---|---|---|---|---|---|
| T1 (n) | $RF^- - RF^+ < 0$ | $-RF^+ < 0$ | | $t0 \rightarrow t3$ | OFF |
| T2 (p) | $RF^- - RF^+ < 0$ | $V_{OUT} - RF^+$ | $<0$ | $t1 \rightarrow t2$ | ON |
| | | | $>0$ | $t0 \rightarrow t1$ $t2 \rightarrow t3$ | OFF, NDR |
| T3 (n) | $RF^+ - RF^- > 0$ | $RF^+ - RF^-$ | $>0$ | $t1 \rightarrow t2$ | ON |
| | | | $<0$ | $t0 \rightarrow t1$ $t2 \rightarrow t3$ | OFF, NDR |
| T4 (p) | $RF^+ - RF^- > 0$ | $V_{OUT} - RF^- > 0$ | | $t0 \rightarrow t3$ | OFF |

**Table 6.2**    Steady-state bias conditions of the TFET-GCCR in region II

| Region II | $V_{GS}$ | $V_{DS}$ | | $\Delta T$ | State |
|---|---|---|---|---|---|
| T1 (n) | $RF^- - RF^+ > 0$ | $-RF^+$ | $>0$ | $t4 \rightarrow t5$ | ON |
| | | | $<0$ | $t3 \rightarrow t4$ $t5 \rightarrow t6$ | OFF, NDR |
| T2 (p) | $RF^- - RF^+ > 0$ | $V_{OUT} - RF^+ > 0$ | | $t3 \rightarrow t6$ | OFF |
| T3 (n) | $RF^+ - RF^- < 0$ | $RF^+ - RF^- < 0$ | | $t3 \rightarrow t6$ | OFF |
| T4 (p) | $RF^+ - RF^- < 0$ | $V_{OUT} - RF^-$ | $<0$ | $t4 \rightarrow t5$ | ON |
| | | | $>0$ | $t3 \rightarrow t4$ $t5 \rightarrow t6$ | OFF, NDR |

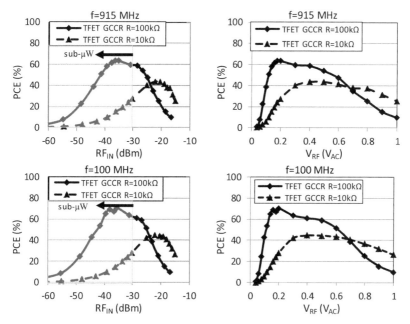

**Figure 6.3** Performance comparison of a TFET-based GCCR considering one stage.

TFETs T1–T4 are simulated with channel widths of 1 μm. For a frequency of 915 MHz (100 MHz), the coupling capacitors CC present 1 pF (10 pF) and the load capacitor $C_L$ 10 pF (100 pF). One can observe that compared to a load of 10 kΩ, a load of 100 kΩ allows larger PCE values at submicrowatt power levels for both frequencies under study. The peak efficiency for a load of 100 kΩ is shown to be around an RF $V_{AC}$ of 0.2 V. At larger voltage magnitudes, the decrease of the PCE is not only due to the increase of conduction losses by the transistors operating in on-state but also due to nonfully closed transistors operating in off-state, and consequent conduction of reverse current.

## 6.3 Advantages of Tunnel FETs in Rectifiers

The improved electrical characteristics of tunneling devices at low-voltage operation in comparison to conventional thermionic devices allow for increased performance of TFET-based rectifiers at ultralow power operation (submicrowatts). As it will be shown in the following sections, when compared to thermionic devices, the lower reverse current and increased drive current at sub-0.25 V characterize TFET-based rectifiers with less reverse and conduction losses and consequent larger PCE at low induced RF voltage. As shown in Tables 6.1 and 6.2, each TFET device operating in a GCCR stage presents during most of the period cycle an off-state condition. As an example, the device T1 is forward biased (on-state) during the time interval $t4 \rightarrow t5$, while during the remaining period time, it is reverse biased, i.e., $V_{DS} < 0$ V. During half of the period cycle ($t0 \rightarrow t3$), the tunneling device T1 is reverse biased with a $V_{GS} < 0$ V, and during the time intervals $t3 \rightarrow t4$ and $t5 \rightarrow t6$, it presents a $V_{GS} > 0$ V. As explained in Section 4.2.2, the reverse current of an n-type tunneling device under this condition ($V_{DS} < 0$ V and $V_{GS} > 0$ V) follows a nonmonotonic characteristic, i.e., the magnitude of the reverse current under low reverse bias increases and then decreases at large reverse bias thus characterizing the TFET with a negative differential resistance (NDR) region. In Tables 6.1 and 6.2, one can observe that this condition applies to all the TFET devices inside the rectifier stage.

In contrast, thermionic devices under such a condition present a reverse current that increases in magnitude with the increase of reverse bias. As shown in Figure 6.4, the comparison between a heterojunction n-TFET

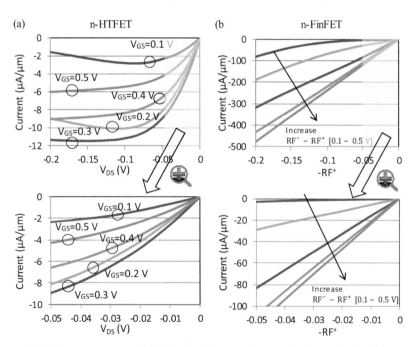

**Figure 6.4** Reverse current of $T1$ during interval $t3 \rightarrow t4$, $t5 \rightarrow t6$ for: (a) n-type heterojunction tunnel field-effect transistor and (b) n-type FinFET.

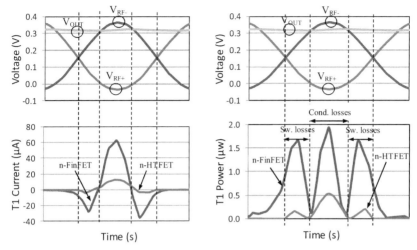

**Figure 6.5** Current and power consumption of transistor T1 in the GCCR during region II of operation. Simulation conditions: $V_{AC}$ = 0.4 V, $f$ = 915 MHz, CC = 1 pF, $C_L$ = 10 pF, and RL = 100 kΩ. WT1-HTFET = 1 μm and NFINS-FinFET = 14.

and thermionic n-FinFET under similar bias conditions (applied during the time intervals $t3 \rightarrow t4$ and $t5 \rightarrow t6$) shows that the reverse current of FinFETs is much larger than that of TFETs. This characteristic results in large losses (called here as switching losses) in FinFET/thermionic-based rectifiers during the mentioned time intervals.

This condition applies to the four transistors in the GCCR during their respective switching time. As an example, Figure 6.5 presents a comparison between the current conducted by the device T1 in the GCCR stage (considering heterojunction n-TFET and n-FinFET) and the respective power consumption when considering an input voltage RF $V_{AC}$ of 0.4 V and a load of 100 kΩ ($f$ = 915 MHz). The switching losses of the FinFET counterpart are shown to be much larger at similar bias conditions, thus resulting in increased losses in FinFET-based rectifiers.

## 6.4 Drawbacks of Tunnel FETs in Rectifiers

As explained in the previous chapter, when the TFET device is subjected to reverse bias conditions, the intrinsic p-i-n diode of its structure is forward biased. With the increase of reverse bias occurring at large induced RF voltages ($V_{DS} < 0$ V in n-TFETs and $V_{DS} > 0$ in p-TFETs), the large increase

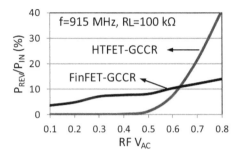

**Figure 6.6**    Increase of reverse losses as a function of RF $V_{AC}$ magnitude.

of reverse current in TFETs strongly degrades the performance of TFET-based rectifiers. As shown in Figure 6.6, the increase of RF voltage magnitude results in an exponential increase of the reverse losses ($P_{REV}$) in the HTFET-GCCR. In contrast, the losses presented by the FinFET counterpart increase only linearly with RF $V_{AC}$ magnitudes. Therefore, this characteristic limits the use of TFETs in rectifiers to low-voltage operation.

## 6.5 Proposed Tunnel FET-based Rectifier

In order to avoid large reverse losses due to large reverse-biased TFETs in rectifiers, a possible solution is to decrease the magnitude of $V_{GS}$ during the off-state condition of the transistor. As explained in the previous chapter and taking as an example the HTFET (GaSb-InAs, $L_G = 40$ nm), the reverse current under a $V_{GS} = 0$ V condition is attenuated at a larger reverse bias range in comparison to a $V_{GS} = V_{DS}$ condition (see Figure 5.3). Therefore, in order to attenuate the reverse current of reverse-biased TFETs in rectifiers, a different topology is required.

In Figure 6.7, a different rectifier topology for TFET devices is proposed. The rectifier is characterized by biasing the gate of the main transistors T1 and T3 with $RF^+$ when $RF^+ > RF^-$ and $RF^-$ otherwise. In a similar way, the gates of transistors T2 and T4 are biased with $RF^-$ when $RF^- < RF^+$ and $RF^+$ otherwise.

In order to accomplish this behavior, transistors T1 and T3 share the same gate, as also T2 and T4. Two auxiliary TFET devices (T5 and T6) are required to bias the gates of T1 and T3 and two auxiliary transistors (T7 and T8) to bias the gates of T2 and T4. In the proposed topology, the $V_{DS}$ bias of the four main transistors (T1–T4) remains the same as the conditions shown in Tables 6.1 and 6.2. In contrast, a different $V_{GS}$ magnitude is applied to the main transistors.

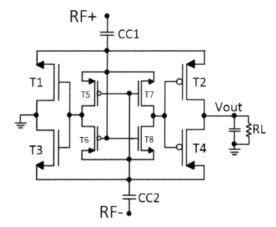

**Figure 6.7**   The proposed tunnel field-effect transistor-based rectifier.

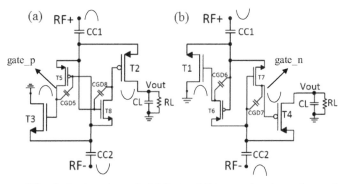

**Figure 6.8**   Active transistors in (a) first and (b) second regions of operation.

Considering an ideal rectifier (no losses in the transistors), during the first region of operation shown in Figure 6.8(a) ($RF^+ > RF^-$), the main transistors T2 and T3 and the auxiliary transistors T5 and T8 are active (T1, T4, T6, and T7 operate in off-state), while the second region of operation shown in Figure 6.8(b) ($RF^+ < RF^-$) characterizes T1, T4, T6, and T7 in on-state (T2, T3, T5, and T8 in off-state). The ideal transient behavior of the proposed rectifier is shown in Figure 6.9.

According to Table 6.3, when the main transistors are reverse biased (highlighted in bold), the magnitude of the gate-to-source voltage $V_{GS}$ is ideally zero (considering the ideal transient behavior shown in Figure 6.9). As shown in the previous chapter, this condition highly reduces the reverse current and consequent reverse losses of the topology. In an ideal case, when

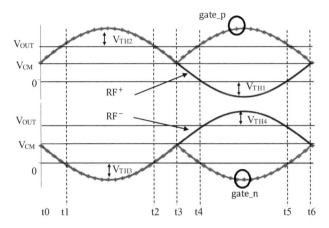

**Figure 6.9**   Ideal transient behavior of the proposed tunnel field effect transistor rectifier.

**Table 6.3**   Ideal $V_{GS}$ conditions of the proposed rectifier in both regions of operation (bold: transistors in reverse bias)

|  | $V_{GS}$ Region I t0 → t3 | $V_{GS}$ Region II t3 → t6 |
|---|---|---|
| T1 (n) | **gate_p−RF$^+$** | gate_p − RF$^+$>0 |
| T2 (p) | gate_n − RF$^+$<0 | **gate_n−RF$^+$** |
| T3 (n) | gate_p − RF$^-$>0 | **gate_p−RF$^-$** |
| T4 (p) | **gate_n−RF$^-$** | gate_n − RF$^-$<0 |

the main transistors are forward biased, their $V_{GS}$ magnitude remains the same as the conventional GCCR topology. At this point, it is important to mention that in the proposed rectifier topology, the auxiliary transistors operating during their off-state condition present a nonzero $V_{GS}$ magnitude, and therefore reverse current in these transistors is expected. In order to mitigate the consequent reverse losses of the auxiliary transistors and improve the PCE of the rectifier stage, one can increase the ratio of widths between the main and auxiliary devices.

## 6.6 Optimization of the Proposed Rectifier

In this section, the proposed rectifier is simulated for two different frequencies (100 and 915 MHz) considering an output load of 100 kΩ. For a frequency of 915 MHz (100 MHz), coupling capacitors of 1 pF (10 pF) and load capacitor of 10 pF (100 pF) are considered. For comparison purposes, the GCCR topology is simulated considering the main transistors T1–T4 as

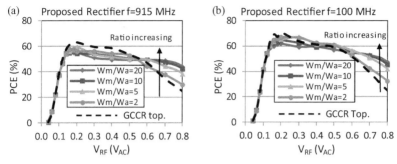

**Figure 6.10** Performance of the proposed rectifier considering (a) $f = 915$ MHz and (b) $f = 100$ MHz.

HTFETs (GaSb-InAs, $L_G = 40$ nm) with widths of 1 µm. In the proposed rectifier, the main transistors are simulated as 1 µm and the auxiliary transistors with different widths.

In Figure 6.10, the performance of the proposed rectifier considering different ratios between the widths of the main (*Wm*) and auxiliary (*Wa*) transistors is presented. One can observe that for both frequencies of operation and RF magnitudes larger than 0.6 $V_{AC}$, the proposed rectifier operates with a larger PCE in comparison to the conventional GCCR (shown in dashed curves). At large voltage, decreasing the size of the auxiliary transistors is shown to increase the rectification efficiency due to the reduction of reverse losses suffered by the auxiliary devices when operating in off-state.

In Figure 6.11, one can observe that the larger efficiency of the proposed rectifier at RF voltages above 0.6 $V_{AC}$ is due to the reduction of the reverse losses of the main transistors during their off-state operation. Despite

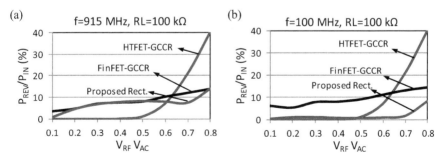

**Figure 6.11** Reverse losses of rectifiers (one stage) as a function of $V_{AC}$ magnitudes considering (a) $f = 915$ MHz (prop. Rect: *Wm/Wa* = 10) and (b) $f = 100$ MHz (prop. Rect: *Wm/Wa* = 5).

the increased performance of the proposed TFET rectifier in comparison to the GCCR counterpart at large voltage, there is a lower efficiency at sub-0.6 $V_{AC}$.

From the circuit shown in Figure 6.8, it can be deduced that the large gate-to-drain intrinsic capacitances ($C_{GD}$) of the auxiliary transistors operating during on-state are responsible for delays between the voltage applied in the gate and the source of the main transistors. As shown in Figure 6.12(a), this characteristic has implications in the transient behavior of the rectifier stage. When considering an operation frequency of 915 MHz, the out-of-phase gate and source voltages fail to provide a $V_{GS}$ close to 0 V when the main transistors are reverse biased. This strongly affects the performance of n-TFETs T1 and T3 under reverse-biased conditions by placing them under an NDR region ($V_{GS} \gg 0$ V and $V_{DS} < 0$ V) and increased conduction of reverse current (see Figure 6.4). In contrast, at a frequency of 100 MHz, the reverse-biased transistors T1 and T3 are placed in an NDR region with a small $V_{GS}$ magnitude and consequent low reverse current ($V_{GS} > 0$ V and $V_{DS} < 0$ V).

Larger auxiliary transistors can mitigate the delays between the gate and source voltages of the main transistors at high frequencies by fastening the charge and discharge rate of their parasitic gate-to-drain capacitances (when forward biased). However, large auxiliary transistors will suffer from large reverse losses at large RF $V_{AC}$ magnitudes when reverse biased. As shown in Figure 6.10, there is a tradeoff between the choice of large auxiliary transistors (increase efficiency at sub-0.6 $V_{AC}$) and small auxiliary transistors (increased efficiency at $V_{AC} > 0.6$ V). In the following section, a ratio

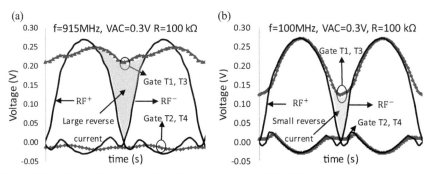

**Figure 6.12**   Gate voltage applied to the main transistors T1–T4 in the proposed rectifier for: (a) $f = 915$ MHz ($Wm/Wa = 10$) and (b) $f = 100$ MHz ($Wm/Wa = 5$).

between the main and auxiliary transistors of 10 and 5 is considered in the proposed rectifier for, respectively, $f = 915$ MHz and $f = 100$ MHz.

## 6.7 Performance Comparison of Rectifiers

In Figure 6.13, a performance comparison between a GCCR and the proposed rectifier designed with HTFETs (GaSb-InAs, $L_G = 40$ nm) is presented. For comparison purposes, the performance of a FinFET-based GCCR is included (FinFETs with $N_{FINs} = 14$ and $L_G = 20$ nm). In both simulations, one rectifier stage is considered.

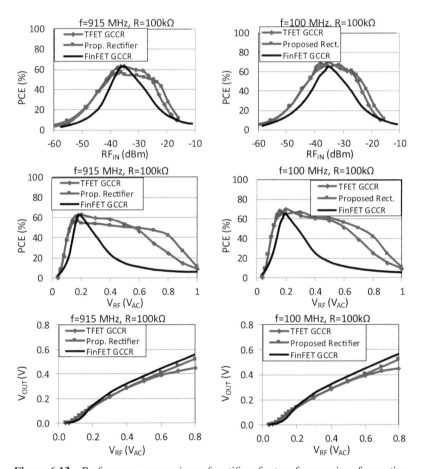

**Figure 6.13** Performance comparison of rectifiers for two frequencies of operation.

**Table 6.4**    Suitable rectifier topology for different voltage/power ranges

|  | Ultra-Low Voltage | | Low-Voltage | | Medium Voltage | |
|---|---|---|---|---|---|---|
| Voltage Range | < 200 mV | | 200–600 mV | | > 600 mV | |
| Frequency | 100 MHz | 915 MHz | 100 MHz | 915 MHz | 100 MHz | 915 MHz |
| Suitable | HTFET-GCCR | HTFET- | Prop. | HTFET- | Prop. | Prop. |
| Topology | Prop. Rect. | GCCR | Rect. | GCCR | Rect. | Rect. |

One can observe that the FinFET-based rectifier produces the lowest PCE at sub-40 dBm when considering the two frequencies of operation under study. It shows, however, the largest output voltage values for RF $V_{AC}$ magnitudes larger than 0.3 V. This characteristic is explained due to the larger current conducted by FinFETs above 0.25 V when compared to HTFETs.

At sub-25 dBm and $f = 915$ MHz, the conventional HTFET-GCCR is shown to produce the largest power efficiencies. At the same frequency, the proposed HTFET-rectifier shows a degraded efficiency at RF power levels between −25 and −40 dBm (corresponding to RF $V_{AC}$ between 0.2 and 0.6 V) due to the increased losses of the main transistors (n-TFETs placed in an NDR region). At sub-40 dBm, the performance of both HTFET-based rectifiers is similar (reverse current of n-TFETs in the proposed rectifier is negligible).

At a frequency of 100 MHz and sub-25 dBm, the proposed HTFET-rectifier and HTFET-GCCR present a similar performance. The proposed HTFET-rectifier shows, however, improved power efficiencies when compared to the counterpart rectifiers at RF power levels above −25 dBm. This is due to the reduction of reverse losses in the main transistors when the proposed HTFET-rectifier is subjected to RF $V_{AC}$ magnitudes larger than 0.6 V. This characteristic is observed for the two frequencies of operation under study.

According to the results, there is suitable rectifier topology for different regions of voltage operation and low/high frequencies. In Table 6.4, a summary of suitable rectifier topologies is presented for different operation conditions.

## 6.8 Chapter Summary

In this chapter, a performance comparison between rectifiers designed with tunneling and thermionic-based devices is presented. It is shown that when the HTFET is reverse biased, the consequent conduction of reverse current at large induced RF voltage strongly degrades the performance of the rectifier stage. Therefore, a different rectifier topology is proposed: when the tunneling devices are reverse biased, a $V_{GS} = 0$ V is applied to their terminals in order to reduce the conduction of reverse current and increase the rectifier stage efficiency. In order to accomplish this characteristic, an auxiliary rectifier inside the rectifier stage is proposed. In order to reduce the losses of the auxiliary circuit, the ratio between the widths of the main and auxiliary transistors can be increased. When compared to the conventional gate cross-coupled topology, the proposed TFET-based rectifier is shown to produce the largest rectification efficiencies at a wider range of voltage/power operation under an operating frequency of 100 MHz.

To conclude, the inclusion of HTFETs in rectifiers can improve the field of RF energy harvesting by allowing larger power conversion efficiencies at power/voltage levels where conventional thermionic devices are shown to be inefficient.

## References

[1] Paing, T., Falkenstein, E. A., Zane, R., and Popovic, Z. "Custom IC for ultralow power RF energy scavenging," in *IEEE Trans. Power Electron.*, vol. 26, no. 6, pp. 1620–1626, 2011.

[2] Adami, S.-E., Vollaire, C., Allard, B., Costa, F., Haboubi, W., and Cirio, L. "Ultra-low power autonomous power management system with effective impedance matching for RF energy harvesting," in *Int. Conf. on Int. Power Syst. (CIPS)*, pp. 1–6, 2014.

[3] Sain, G., Arrawatia, M., Sarkar, S., and Baghini, M. S. "A battery-less power management circuit for RF energy harvesting with input voltage regulation and synchronous rectification," in *Int. Mid. Symp. on Circ, and Syst.*, pp. 1–4, 2015.

[4] Jinpeng, S., Xin'an, W., Shan, L., Hongqiang, Z., Jinfeng, H., Xin, Y., et al. "Design and implementation of an ultra-low power passive UHF RFID tag", *J. Semicond.* 33:115011, 2012.

[5] Liu D.-S., Li, F. B., Zou, X. C., Liu, Y., Hui, X. M., and Tao, X. F. "New analysis and design of a RF rectifier for RFID and implantable devices," *Sensors* 11, 6494–6508, 2011.

[6] Jinpeg, S., Xin'an, W., Shan, L., Hongqiang, Z., Jinfeng, H., Xin, Y., et al. "A passive RFID tag with a dynamic-VTH-cancellation rectifier," *J. Semicond.* 34:95005, 2013.

[7] Burasa, P., Constantin, N. G., and Wu, K. "High-efficiency wideband rectifier for single-chip batteryless active millimeter-wave identification (MMID) tag in 65-nm Bulk CMOS technology," in *IEEE Trans. onMicrowave Theory and Techniques*, vol. 62, no. 4, pp. 1005, 1011, 2014.

[8] Mandal, S., and Sarpeshkar, R. "Low-power CMOS rectifier design for RFID applications," *IEEE Trans on Circuits Syst.* vol. 54, no. 6, pp. 1177–1188, 2007.

[9] Liu, H., Li, X., Vaddi, R., Ma, K., Datta, S., and Narayanan, V. "Tunnel FET RF rectifier design for energy harvesting applications," *IEEE Journal on Emerging and Selected Topics in Circuits and Systems*, vol. 4, no. 4, pp. 400–411, 2014.

# 7

# TFET-based Power-management Circuit for RF Energy Harvesting

In this chapter, a Tunnel Field Effect Transistor (TFET)-based Power-management Circuit (PMC) for ultra-low power RF energy harvesting (EH) applications is proposed. The advantages of using tunneling devices in RF PMC are identified, as also the challenges on the design of inductor-based boost converters due to the different electrical characteristics of TFETs under reverse-bias conditions. The proposed TFET-PMC shows promising results at available RF power levels below $-20$ dBm (f = 915 MHz). For an available power of $-25$ dBm, the proposed converter is able to deliver 1.1 $\mu$W of average power to a load (0.5 V) with a boost efficiency of 86%.

In order to allow maximum power transfer between the front-end rectifier and the boost converter, the TFET-PMC adapts its input impedance. Once the output of the boost converter reaches 0.5 V, a load is enabled and the PMC starts a self-sustaining mode (SSM) of operation. All the simulated results are based on the Verilog-A based LUT TFET models and PTM-based FinFET models described in Chapter 4.

## 7.1 Motivation

With the fast progression in the development of low-power embedded systems, the design of efficient circuits at reduced voltage operation has gained momentum in recent years [1–2]. Low voltage and power systems, like biomedical implants or wearable devices are examples of applications that can benefit from harvesting ambient electromagnetic radiation, thus reducing battery size and extending its lifetime. Several works have already demonstrated wireless powering of a load at short distances with UHF radiation at legally transmitted power levels [3–11]. However, the received radiation power attenuates with distance and the low power-conversion efficiency (PCE) demonstrated by front-end rectifiers at low RF power levels (sub-20 dBm) constrains the operation distance of RF energy harvesters to short distances.

The low power levels of electromagnetic radiation produce low output voltage values in the receiving antenna and therefore, efficient rectifiers are required for proper system operation. In Chapter 6, it was shown by simulation that TFET-based rectifiers can outperform thermionic-device-based counterparts in extreme low-voltage/power scenarios (sub-20 dBm). In addition and as shown in Section 4.3, the particular electrical characteristics of tunneling devices enable the decrease of energy per switch operation, thus allowing the design of efficient digital cells at low voltage in comparison to conventional CMOS [12–14]. This characteristic allows the design of efficient PMC at low voltage operation (sub-0.25 V). Therefore, it is of interest to explore the performance and limitations of tunneling devices in ultra-low voltage PMCs for RF EH applications.

## 7.2 Challenges in RF Power Transport

In Figure 7.1, the structure of the RF power-transport system considered in this work is presented. The receiver comprises a receiving antenna with 50 $\Omega$ standard impedance followed by a lumped matching network between the antenna and the rectifier. The PMC is required to boost the rectifier output voltage to a larger and stable voltage in order to power a load/sensor.

The main challenge in the receiver of the RF power system is to overcome the power-density attenuation due to the long distances between the transmitter and receiver circuits. The Friis equation expressed by (7.1) indicates that the power received at the input of the rectifier $P_R$ is dependent on the transmitted power $P_T$, transmitter and receiver antenna gains $G_T$ and $G_R$ respectively, wavelength $\lambda_o$ of the transmitter signal, and propagation

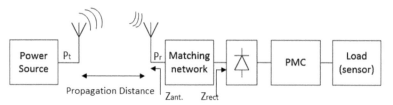

**Figure 7.1**    RF Power Transport System.

distance R [1]. If the receiver antenna is well-matched with the rectifier, the relation between the peak amplitude of the antenna $V_A$ and the received power can be expressed by Equation (7.2), with $R_A$ representing the real part of the antenna impedance.

$$PR = P_T G_T G_R \left( \frac{\lambda_o}{4\pi R} \right)^2 \qquad (7.1)$$

$$V_A = \sqrt{8R_A P_r} \qquad (7.2)$$

In Table 7.1, the license-free Industry-Science-Medical (ISM) frequency bands for different regions are indicated [2]. Considering two different frequency bands (915 MHz and 2.4 GHz), a maximum regulated transmitter power of 4 W, and taking into account Equations (7.1) and (7.2), one can calculate the received power and the peak amplitude of the antenna as a function of the propagation distance R as shown in Figure 7.2 (assuming antenna gain of 1). It is shown that the power density attenuation at the input of the rectifier as a function of the propagation distance constrains the operation of RF systems to short distances.

As an example and according to Figure 7.2(a), the receiving antenna senses −25 dBm at a propagation distance of 30 m from the transmitter (f = 915 MHz, $P_T$ = 4 W). As shown in Figure 7.2(b), at 30 m, the receiving

**Table 7.1** Frequency Band allocations and maximum transmitter power [1]

| Freq. Band (MHz) | Transmitter Power | Region |
|---|---|---|
| 2446–2454 | 500 mW–4 W (EIRP) | Europe |
| 867.6–868 | 500 mW (ERP) | Europe |
| 902–928 | 4 W (EIRP) | USA/Canada |
| 2400–2483.5 | 4 W (EIRP) | USA/Canada |
| 2400–2483.5 | 10 mW (EIRP) | Japan/Korea |

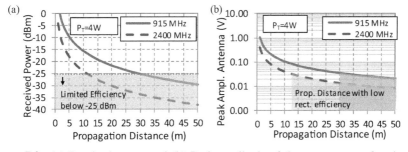

**Figure 7.2** (a) Received power and (b) Peak amplitude of the antenna as a function of propagation distance for $P_T$ = 4 W.

antenna produces a peak voltage amplitude of 36 mV. At such low voltage levels, there is a clear difficulty of rectification as shown by the recent results from the state-of-the-art CMOS rectifiers shown in Figure 7.3.

In order to increase the input voltage of the rectifier and consequent PCE of rectification, a resonating LC network optimized for a desired input power level can be included at the output of the antenna as shown in Figure 7.4(a) [1, 6]. This method is adopted in the proposed RF PMC. In the front-end part of the RF PMC, the gate cross-coupled rectifier (GCCR) topology shown in Figure 7.4(b) is considered and matched with a 50 $\Omega$ antenna.

In Figure 7.5(a), the PCE of a GCCR with one-stage designed with different technologies is presented. For the rectifier design, heterojunction TFETs (InAs-GaSb, $L_G$ = 40 nm), homojunction TFETs (InAs, $L_G$ = 20 nm), and Si-FinFETs ($L_G$ = 20 nm) are considered. Each transistor is simulated with a channel width of 10 $\mu$m. The matching network with $L$ and $C$ elements is optimized for each rectifier considering an available power of $-20$ dBm.

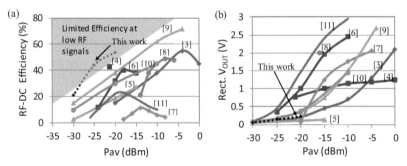

**Figure 7.3** Comparison of (a) CMOS rectifiers from literature and (b) Rectifier output voltage as function of available power.

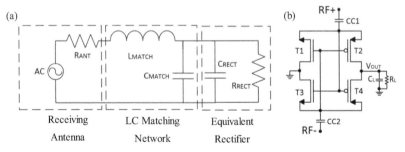

**Figure 7.4** (a) Equivalent circuit of antenna-matching-rectifier and (b) TFET-based GCCR topology.

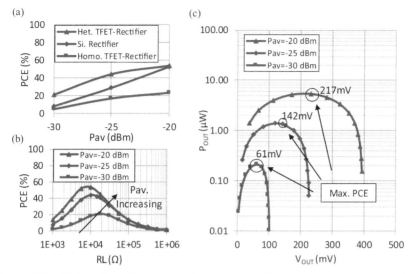

**Figure 7.5** (a) Rectifier efficiency as function of available power considering different technologies; (b) Heterojunction TFET-based rectifier PCE as a function of output load; (c) Rectifier output power as a function of output voltage for heterojunction TFET-based rectifier.

The results show that the heterojunction TFET-based rectifier presents higher rectification efficiencies in the range of −30 to −25 dBm when compared to the FinFET-based rectifier. At such low power levels the homo-junction TFET-based GCCR presents the lowest efficiency values. Compared to the FinFET-based rectifier, the higher efficiency shown by the hetero-junction TFET-based counterpart is explained due to the better switching performance of individual TFET transistors at sub-0.2 V operation (see Chapter 6).

With a focus on the heterojunction TFET-based rectifier, Figure 7.5(b) shows that the PCE changes according to the load. One can observe that increasing the available power requires lower load values for increased efficiencies. As shown in Figure 7.5(c), increasing the available RF power increases the output voltage of the rectifier where maximum efficiency is achieved. For a good RF PMC design, knowing the load range that maximizes the rectifier efficiency in advance is mandatory in order to allow maximum power transfer from the rectifier to the input of the boost converter.

## 7.3 Proposed TFET-based PMC

In Figure 7.6, the building blocks of the proposed RF TFET-based PMC are presented. The RF power source is simulated as a 50 $\Omega$/915 MHz port. After matching and rectification the PMC is required to boost the low output voltage of the rectifier to 0.5 V and then enable a load. The PMC is divided into three distinct modules: startup, controller and boost circuit.

The first module is responsible for pre-charging the power capacitors connected to nodes $VDD_{INT}$ and $VDD_{STARTUP}$ (by the rectifier) and the input $C_{BOOST}$ and output $C_{OUT}$ capacitors of the boost converter circuit to adequate voltage values before enabling the boost-converter operation.

The controller circuit (powered by the capacitors of the startup module) is responsible for providing control signals to the switches of the boost-converter circuit. When the boost converter operation takes place and the load is enabled for the first time, the PMC enters into an of operation, i.e. the power capacitors are directly charged by the output capacitor $C_{OUT}$ and

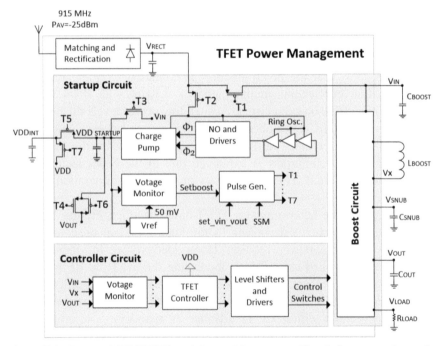

**Figure 7.6**  Proposed RF TFET-based Power Management Circuit for energy harvesting applications.

not by the rectifier. In order to increase the PMC efficiency, the controller module is responsible for imposing an adequate voltage to the input of the boost converter (that interfaces with the output of the rectifier) in order to maximize the rectifier efficiency.

### 7.3.1 Startup Circuit

The proposed startup module is designed with the purpose of avoiding the use of any external power source for a proper PMC operation. As shown in Figure 7.6, a ring oscillator powered by the rectifier is required for generating two non-overlapped clock signals that are applied to a multi-stage gate cross-coupled charge pump (GCCCP). In Chapter 5, it was shown by simulations that a TFET-based GCCCP (designed with heterojunction TFETs) can double the input voltage with magnitudes as low as 80 mV. In this work, the charge pump is required to charge the capacitor connected at node $VDD_{STARTUP}$ in order to power all the analog and digital circuitry of the startup module.

As shown in Figure 7.7, a voltage monitor is required to trigger a signal *Setboos*t each time the node $VDD_{STARTUP}$ reaches 200 mV. This signal is

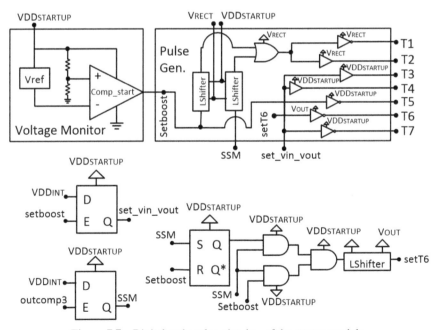

**Figure 7.7** Digital and analog circuitry of the startup module.

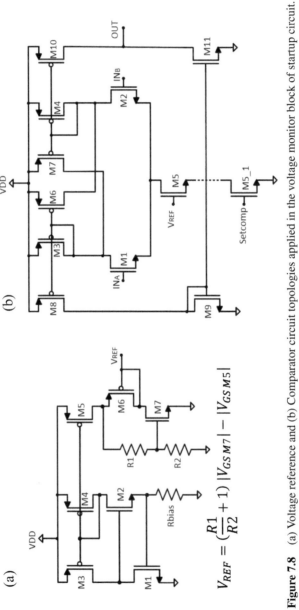

**Figure 7.8**    (a) Voltage reference and (b) Comparator circuit topologies applied in the voltage monitor block of startup circuit.

required to enable the boost conversion. A voltage reference circuit such as the one shown in Figure 7.8(a) provides a fixed 50 mV reference to the comparator of the voltage monitor block. In [15] the authors have shown that this voltage reference topology designed with TFETs presents a low dependence on power supply voltage and temperature. The comparator of the voltage monitor is based on the hysteresis comparator shown in Figure 7.8(b). The 50 mV of the voltage reference circuit is applied to the negative input of the comparator (M2) and also required to bias the differential pair (through M5).

Before enabling the boost conversion, the input $C_{BOOST}$ and output $C_{OUT}$ capacitors of the boost converter are pre-charged to 200 mV (from node $VDD_{STARTUP}$) by the TFET switches controlled by T3 and T4. Once charged, a signal *set_vin_vout* is enabled and the boost operation starts. The TFET switch controlled by T1 is required to allow the charging of the capacitor connected to the input of the boost converter. The output of the rectifier is responsible for this charging each time the signal *Setboost* is active. The TFET switch controlled by T5 is required to allow the charging of the capacitor at node $VDD_{INT}$ to the same voltage level of node $VDD_{STARTUP}$, i.e. 200 mV and the switch controlled by T7 to enable this capacitor as the power source of the digital and analog circuitry in the controller module.

Once the output load of the system is enabled, i.e. node $V_{LOAD}$ goes from low-to-high state, a signal SSM is triggered and the TFET switch controlled by T2 deactivates the ring oscillator, clock signals, and the charge pump circuit. Then, the output capacitor $C_{OUT}$ is responsible for charging the capacitor connected to the node $VDD_{STARTUP}$ through the TFET switch controlled by T6. In the digital circuitry of the startup module, the signal *outcomp3* (coming from the controller module) is responsible for triggering the SSM signal. Inside the startup module, two level-shifter circuits are required to match the voltage of both the input and power source of digital cells. The relevance of level shifters (LSs) in TFET-based circuits is explained in Section 7.3.3.

## 7.3.2 Boost Circuit

In order to increase the output voltage of the rectifier to voltage values suitable to be applied to the output load, a boost converter is required. In Figure 7.9, an inductor-based boost converter topology for TFET devices is proposed. The sequence of operation for the gate controls of each TFET device is shown in Figure 7.10.

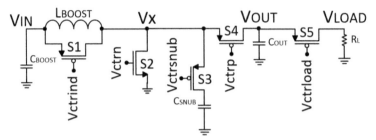

**Figure 7.9**    Proposed boost converter topology for TFET devices.

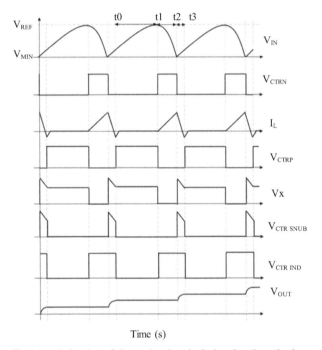

**Figure 7.10**    Transient behavior of the main electrical signals when the boost controller is enabled.

After a proper startup operation and *Setboost* signal is enabled, the input capacitor $C_{BOOST}$ is charged and discharged maintaining an average voltage (matching voltage) adequate for maximum rectifier efficiency. This matching technique is used by several works [16–18], allowing for maximum transfer of power from the rectifier to the boost converter.

The suitable matching voltage depends on the received power. According to the rectifier output power as a function of output voltage shown in Figure 7.5(c), the optimum reference voltage at the input of the boost converter (output of rectifier) for maximum power efficiency is 142 mV if an available power of –25 dBm is considered. If the available RF power changes, the reference voltage value at the input of the boost converter has to change accordingly. In order to perform a rectifier-boost matching, the PMC charges and discharges $C_{BOOST}$, maintaining an average voltage close to 142 mV ($V_{IN}$ between $V_{REF}$ and $V_{MIN}$ as shown in Figure 7.10).

During the time interval t0 to t1, the input capacitor $C_{BOOST}$ is charged by the rectifier up to $V_{REF}$. During this time interval, no current should flow through the inductor. In order to accomplish this, the TFET device S1 in Figure 7.9 is closed and a small current flows through the inductor.

The absence of body diode in reverse-biased TFETs (due to a different doping structure than that of thermionic MOSFETs) requires a change in the conventional boost converter topology. During the reverse bias time interval of the output transistor S4 (t0 to t1) a snubber circuit (designed with TFET device S3 and a capacitor $C_{SNUB}$) provides a path for the inductor current to flow. During the time interval t1 to t2, the TFET device S2 is closed in order to charge the inductor from the input capacitor, until the latter is discharged to a minimum reference value $V_{MIN}$. During this time interval the remaining transistors are open. During the time interval t2 to t3, the TFET S4 is closed in order to allow the transfer of stored energy in the inductor to the output capacitor $C_{OUT}$, thus increasing the output voltage of the boost converter. When the output of the boost converter achieves a reference value of 510 mV, the TFET device S5 is closed and an external load is enabled until the output capacitor is discharged to 490 mV.

### 7.3.2.1 Challenges in TFET-based boost-converter design

In order to increase the PCE of inductor-based boost converters with TFETs, it is mandatory to keep a low reverse-current conduction from the output transistor S4 during the time interval t0 to t2, i.e., when the device is reverse biased ($V_{DS} > 0$ V) [see Figure 7.11(a)]. In the previous chapters, it was shown that reducing the gate bias magnitude of reverse-biased TFETs allows for the reduction of reverse current at low reverse drain bias. However, at large reverse drain bias, the reverse current of TFETs follows the behavior of a diode. This characteristic is shown in Figure 7.11(b), i.e. the increase of drain to source voltage magnitude in the p-TFET device S4 results in

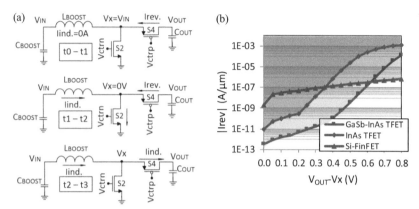

**Figure 7.11**    (a) Top: $C_{BOOST}$ charging from rectifier; Middle: Inductor charging from $C_{BOOST}$; Bottom: Inductor discharging to $C_{OUT}$; (b) Reverse current for different technologies as a function of $V_{DS}$ in S4.

increased reverse current and a consequent increase of discharge rate in the output capacitor. The reverse current is simulated for three technologies (homojunction and heterojunction TFETs and Si-FinFET) as a function of $V_{DS}$ ($> 0$ V, p-device S4 reverse biased). The results consider TFETs with a $V_{GS}$ magnitude equal to 0 V, while a gate voltage equal to $V_{OUT}$ is considered for the reverse-biased FinFET.

During the time interval t0 to t1, the source node of the output transistor S4 equals $V_{IN}$ (assuming that no current is flowing through the inductor) and therefore $V_{DS}$ of S4 equals $V_{OUT}-V_{IN}$. As an example, if the input voltage of the boost converter is 0.1 V and the required output voltage is 0.8 V ($V_{DS} = 0.7$ V), then during this time interval the reverse current conducted by the GaSb-InAs TFET device S4 is approximately 2 orders of magnitude larger than that of the FinFET device S4 for a similar channel width. Since the time interval t0 to t1 dominates the period of the boost operation, an increase of boost frequency can reduce the discharge rate of the output capacitor $C_{OUT}$.

During the time interval t1 to t2, the source node of the output transistor S4 is at 0 V and its $V_{DS}$ equals $V_{OUT}$. Considering the previous example, if one requires an output voltage of 0.8 V, then the reverse current conducted by the heterojunction TFET device S4 is more than 2 orders of magnitude larger than that of the FinFET device S4 when considering a similar channel width. During this time interval, an inductor current larger than the reverse current of S4 is required in order to allow the storage of energy in the inductor. In order to reduce the discharge rate of $C_{OUT}$ and consequently improve the boost efficiency, several solutions can be adopted: decrease the boost frequency (requires larger inductor current), decrease the size of S4 or increase the size of the output capacitor.

During the time interval t2 to t3, S4 is a forward-biased conducting current from the inductor to the output capacitor and therefore, a large output transistor channel width is preferred in order to attenuate the conduction losses. Therefore, there is a trade-off between choosing a large transistor S4 to reduce forward conduction losses or a small transistor channel to attenuate the reverse current.

### 7.3.2.2 Advantages of TFETs in PMC and boost converters

The analysis of the internal resistance of a transistor as a function of $|V_{DS}|$ is useful to evaluate the performance of the device in a PMC and boost converter. As shown in Figure 7.12, the heterojunction TFET device presents the lowest internal resistance under forward bias conditions at sub-0.25 V. When compared to conventional MOSFETs, this characteristic allows for decreased conduction losses in the input transistor S2 during the time interval t1 to t2, and decreased conduction losses in the output transistor T4 during the charge time of the output capacitor (time interval t2 to t3). Furthermore, the lower static and dynamic power consumption of TFET-based circuits at 0.2 V allows for a minimization of the energy required for proper boost controller operation when compared to the use of conventional thermionic technologies. The larger current conducted by TFETs at sub-0.2 V operation also enable

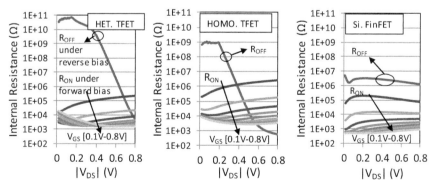

**Figure 7.12**    Internal resistance of different technologies (n-type) under reverse ($V_{DS} < 0$ V) and forward ($V_{DS} > 0$ V) bias. FinFETs under reverse bias present $V_{GS} = V_{DS}$, whereas for TFET $V_{GS} = 0$ V.

the design of buffers (applied to the transistors of the boost converter) with smaller dimensions.

In comparison to Si-FinFETs and under a specific reverse-bias condition range, the larger internal resistance of Tunnel FETs ($R_{OFF}$) allows for reduced reverse losses in boost converters. As shown in the previous section, this characteristic is presented as an advantage if the output transistor of the boost converter (S4) presents a low reverse-bias magnitude, e.g. $|V_{DS}| <$ 0.3 V for homojunction TFET and $|V_{DS}| < 0.6$ V for heterojunction TFET. At larger reverse-bias magnitudes, the reverse current conducted by TFETs (larger than Si-FinFETs) increases the reverse losses of the boost converter, constraining the operation of TFET-based boost converters to low voltage operation.

### 7.3.3 Controller Circuit

For a proper boost-converter operation, a controller is required. In Figure 7.13, a TFET-based controller is proposed. In order to reduce the reverse losses of the TFETs and improve the controller efficiency, the circuit imposes a $V_{GS} = 0$ V to all the TFET devices under reverse bias state. The differential pairs of the two-stage comparators are biased with 50 mV [see Figure 7.8(b)] coming from the voltage reference of the startup circuit.

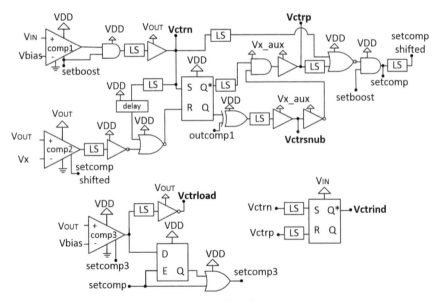

**Figure 7.13**   Proposed TFET-based controller circuit applied to the boost converter.

The first comparator is required to trigger the *Vctrn* signal (applied to S2), maintaining the voltage of the input capacitor $C_{BOOST}$ between a minimum ($V_{MIN}$) and a reference voltage ($V_{REF}$). The second comparator is required to detect a change of polarity in the inductor current, triggering a *Reset* signal that is applied to an RS latch. Depending on the state of *Vctrn* the output transistor S4 is conducting or blocking current according to the signal *Vctrp*. The third comparator is required to enable/disable the output load RL by controlling the TFET device S5. The load is enabled when the output voltage $V_{OUT}$ is between the range of 490 mV and 510 mV. To maximize the controller efficiency, the second and third comparators (*comp2* and *comp3*) only operate during the fraction of time when both *Vctrn* and *Vctrp* signals are at 0 V. This condition triggers the signal *setcomp* that enable the differential pair of the comparator as shown in Figure 7.8(b). The first comparator is enabled as long as the signal *Setboost* is active.

In order to reduce the current through the inductor during the time interval t0 to t1 the control signal *Vctrind* is triggered from an RS latch when both the input S2 and output S4 transistors are operating during their off-state. The *Vctrp* and *Vctrsnub* signals are generated by buffers powered by *Vx_aux*. Since the node *Vx* of Figure 7.11(b) is grounded during the operation time interval t1 to t2, the mentioned buffers cannot be powered directly by *Vx*. In Figure 7.14(a), the proposed circuit guarantees that during the time interval t1 to t2, the voltage node *Vx_aux* equals the voltage node $V_{IN}$ (M1 is open and M2 closed). During the time interval t2 to t3, the node *Vx* is at a higher voltage than that of the node $V_{IN}$ and therefore *Vx_aux* equals *Vx* (M1 is closed and M2 is open).

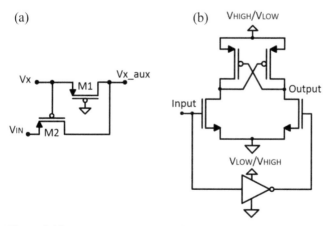

**Figure 7.14**  (a) Vx_aux generator circuit and (b) LS circuit block.

TFET-based digital cells are very sensitive to mismatches between digital levels and power supply. In order to improve the controller efficiency, LS blocks are required to match the voltage at the input of the various digital cells with the applied power-supply voltages. In Figure 7.14(b) the topology of the TFET-based LS is presented. A detailed study on TFET-based LS and the power consumption associated with the voltage conversion can be found in [19].

Figure 7.15 shows the increase of current consumption in several digital cells used in the proposed PMC as a function of ratio between the input voltage of the cell and a power supply voltage of 0.2 V. In the $y$ axis, the current gain is calculated as $I_{MATCHED}*/I_{MATCHED}$ where $I_{MATCHED}$ is the nominal current consumption of the cell when the magnitude of the input voltage of the cell equals the power supply voltage (0.2 V) and $I_{MATCHED}*$ is the current consumption of the cell when the input voltage of the cell is lower than that of the power supply voltage. In Figure 7.16, the increase of current consumption is shown for inverter cells.

From both simulations, one can observe that the current consumption (and consequent power consumption) of TFET-based digital cells is much more sensitive to mismatches between input and power-supply voltages than that of thermionic-based digital cells. This behavior is directly related to the exponential increase of current in TFETs (at low voltage) due to the different carrier injection mechanism. Therefore, in order to avoid an exponential increase of power consumption in TFET-based cells due to voltage mismatches and consequent degradation of power efficiency in the proposed PMC, the use of LS is mandatory.

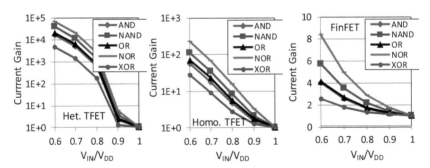

**Figure 7.15**    Increase of current consumption in digital cells for non-matched input ($V_{IN}$) and power supply voltage ($V_{DD}$ = 0.2 V).

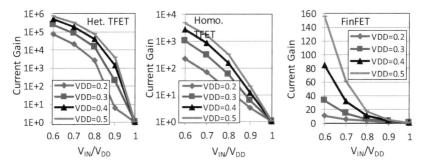

**Figure 7.16** Increase of current consumption in inverter cells for non-matched input ($V_{IN}$) and power supply voltage.

## 7.4 Simulation Results

This section presents simulation results that explore the performance of the proposed PMC designed with heterojunction III–V TFETs (InAs-GaSb, $L_G = 40$ nm) based on the Verilog-A LUT models described in Chapter 4. In Figure 7.17, the transient behavior of the circuit is presented for an available RF power of –25 dBm. It is shown that prior to the boost conversion operation, the input $C_{BOOST}$ and output $C_{OUT}$ capacitors are pre-charged to 200 mV. Once charged, the power supply node of the controller is enabled (VDD) and the boost converter starts a synchronous mode of operation.

One can observe that once the load is enabled, the circuit enters an SSM of operation, i.e., the output capacitor is responsible for charging the capacitors required to power the startup module and the controller ($C_{VDDINT}$ and $C_{VDDSTARTUP}$). With SSM active, the ring oscillator and charge-pump are deactivated, and the charge rate of the capacitors presented in the startup module is faster, thus reducing the off-time of the boost conversion (*Setboost* signal off).

The voltage at the input node of the boost converter ($V_{IN}$) is regulated with an average voltage of approximately 142 mV, thus allowing maximum transfer of power from the rectifier (1.28 µW) according to Figure 7.5(c). During the time interval t1 to t2, $V_{IN}$ and the inductor current are related to the inductor $L$ and input capacitor $C_{BOOST}$ as follows [18]:

$$V_{IN}(t) = V_{REF}\cos\left(\frac{t}{\sqrt{L \times C_{BOOST}}}\right), \quad t_1 \to t_2 \tag{7.3}$$

$$i(t) = V_{REF}\sqrt{\frac{C_{BOOST}}{L}}\sin\left(\frac{t}{\sqrt{L \times C_{BOOST}}}\right), \quad t_1 \to t_2 \tag{7.4}$$

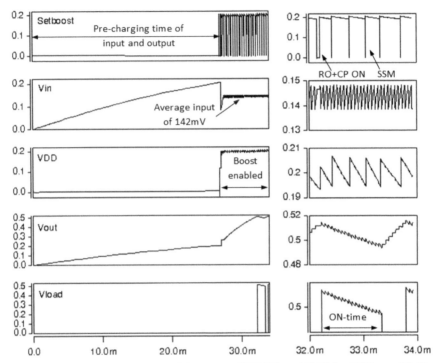

**Figure 7.17** Transient simulation of the proposed TFET-based PMC for RF Pav = –25 dBm. L1 = 10 mH, $C_{BOOST}$ = 0.05 µF, $C_{OUT}$ = 0.05 µF, $C_{SNUB}$ = 2 nF, RL = 166.7 kΩ, WS1 = 10 µm, WS2,3,5 = 100 µm, WS4 = 25 µm.

According to the previous expressions, the inductor current and the boost frequency are proportional to the capacitance value $C_{BOOST}$.

Larger current values due to larger capacitances require input and output transistors with larger channel widths in order to reduce the forward losses and increase the PCE of the boost convertor. Whereas the increase of the output transistor size can reduce its forward losses, the increase of the reverse current conduction and consequent reverse losses degrades the PCE of the boost converter.

In Figure 7.18 the performance of the boost converter is shown considering an output load RL of 166.7 kΩ. Once the load is enabled, an instantaneous output power of 1.5 µW is observed. When considering an input boost capacitor with 0.05 µF and inductor with 10 mH, boost conversion efficiencies close to 90% are achieved. One can observe the presence of an optimum channel width in the output transistor S4 that maximizes the efficiency of the boost

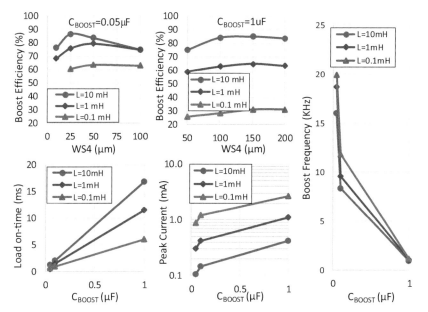

**Figure 7.18** Performance of TFET-based boost converter for RF Pav = −25 dBm. $C_{OUT} = C_{BOOST}$, $C_{SNUB}$ = 2 nF, RL = 166.7 kΩ, WS1 = 10 μm, WS2,3,5 = 100 μm.

conversion. This value is dependent on the inductor size and input capacitance of the boost converter. It is also shown that despite the consequent increase of the circuit die area, the choice of large inductors produces larger conversion efficiency values due to the consequent reduction of the peak current in the boost converter (see Equation 7.4).

As shown in Figure 7.18, low inductor values result in large inductor currents and consequent losses in the switches of the boost converter. For an inductor with 10 mH, a parasitic series resistance of 30 Ω is considered. As the average inductor current is in the order of μA, the losses associated with this resistance value represent a small fraction of the total power losses in the boost converter.

Figure 7.19 presents the power losses distribution of the TFET-based PMC where maximum boost conversion efficiency is achieved: PCE = 86%, WS4 = 25 μm, $C_{OUT} = C_{BOOST} = 0.05$ μF, L = 10 mH. The startup circuit is shown to consume 41.9 nW, the controller circuit, 11.88 nW, and the boost converter, 116 nW, with a great part of the losses resultant from the TFET switches S1 to S5. It is shown that the input and output transistors S2 and S4 respectively are responsible for more than 85% of the boost converter losses (S2 due to forward losses and S4 due to reverse losses). Despite such losses, the performance of the proposed TFET-based PMC shows promising results for μW applications in comparison to the recent RF-PMC from

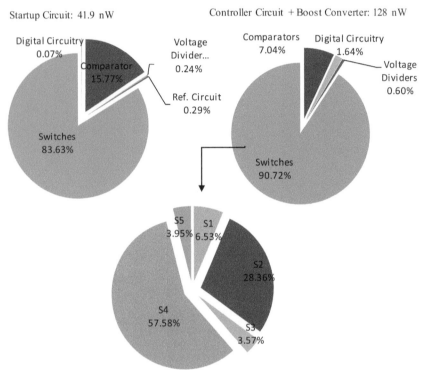

**Figure 7.19** Distribution of power losses in the PMC for RF Pav = −25 dBm. L1 = 10 mH, $C_{BOOST}$ = 0.05 μF, $C_{OUT}$ = 0.05 μF, $C_{SNUB}$ = 2 nF, RL = 166.7 kΩ, WS1 = 10 μm, WS2,3,5 = 100 μm, WS4 = 25 μm.

**Table 7.2** Performance comparison of the proposed TFET-PMC with state of the art

|  | [16] | [17] | [18] | This work |
|---|---|---|---|---|
| RF Freq. | 1.93 GHZ | 2.45 GHz | 950 MHz | 915 MHz |
| Tech. | 350 nm | – | 180 nm | 40 nm TFET |
| Startup | Ext. Battery | Battery-less | Battery-less | Battery-less |
| $V_{OUT}$ | 1.4 V | 1 V | 1 V | 0.5 V |
| $\overline{P_{OUT}}$ | 0.52 μW | 5 μW | 13.1 μW | 1.1 μW |
| PCE DC–DC | 35.13% | 50% | 80% | 86% |
| PCE RF–DC | 0.87% | 15.8% | 13% | 34.8 |
|  | @−12.26 dBm | @−15 dBm | @−10 dBm | −25 dBm |

the literature (shown in Table 7.2). The low power consumption presented by the startup and controller circuits allow for large DC–DC conversion efficiencies at low voltage/power operation whereas the good rectification performance at −25 dBm (approx. 40%) is related to the improved electrical characteristics of TFETs at sub-0.25 V (see Chapter 6) when compared to conventional thermionic devices.

## 7.5 Chapter Summary

In comparison to conventional thermionic devices such as Si-FinFETs, the better switching performance of heterojunction TFETs (InAs-GaSb, $L_G$ = 40 nm) at sub-0.25 V allows for efficient rectification performance at available RF power levels below –20 dBm. It is shown by simulation that at –25 dBm, a TFET-based PMC can boost the output voltage of the rectifier (140 mV) to 500 mV with high efficiency. A TFET-based startup, controller circuit, and boost converter are proposed with power-consumption values of 41.9 nW, 11.88 nW, and 116 nW respectively. These values allow the design of an efficient EH system, showing high boost-conversion efficiencies at input power levels in the μW range.

Reverse current in reverse-biased TFETs present a challenge in the design of TFET-based boost converters when compared to conventional thermionic technologies. Boost converters with larger output values require larger peak inductor currents to counteract the reverse current conducted by the TFET output transistor. This imposes a limitation in the application of TFETs in inductor-based boost converters, i.e., larger output-voltage values in the boost-converter results in larger reverse losses in the output transistor.

The reduction of the $V_{GS}$ magnitude in reverse-biased TFETs (intrinsic *p-i-n* diode forward biased) is shown as a good practice to attenuate the reverse power losses in TFET-based circuits. In order to increase the RF-powered system efficiency, the proposed PMC imposes $V_{GS}$ = 0V to all the Tunneling FET devices under reverse-bias conditions.

Although the presented results do not include pad connection losses and parasitics, the improved switching performance shown by the TFET models when compared to similar device models of thermionic transistors demonstrates the potential of using III-V TFET devices in RF EH applications at μW power levels.

## References

[1] Visser, H. J., and Vullers, R. J. M. "RF energy harvesting and transport for wireless sensor network applications: principles and requirements," in *Proc. IEEE*, vol. 101, no. 6, pp. 1410–1423, 2013.

[2] Soyata, T., Copeland, L., and Heinzelman, W. "RF energy harvesting for embedded systems: a survey of tradeoffs and methodology," in *IEEE Circuits and Syst. Magazine*, vol. 16, no. 1, pp. 22–57, 2016.

[3] Marian, V., Allard, B., Vollaire, C., and Verdier, J. "Strategy for microwave energy harvesting from ambient field or a feeding source," in *IEEE Transactions on Power Electronics*, vol. 27, no. 11, pp. 4481–4491, 2012.

[4] Yao, C., and Hsia, W. "A 21.2 dBm dual-channel UHF passive CMOS RFID tag design," in *IEEE Trans. Circuits Syst. I*, vol. 61, no. 4, pp. 1269–1279, 2014.

[5] Mikeka, C., Arai, H., Georgiadis, A., and Collado, A. "DTV band micropower RF energy-harvesting circuit architecture and performance analysis," *2011 IEEE International Conference on RFID-Technologies and Applications (RFID-TA)*, pp. 561–567, 2011.

[6] Stoopman, M., Keyrouz, S., Visser, H. J., Philips, K., and Serdijn, W. A. "Co-design of a CMOS rectifier and small loop antenna for highly sensitive RF energy harvesters," in *IEEE Journal of Solid-State Circuits*, vol. 49, no. 3, pp. 622–634, 2014.

[7] Papotto, G., Carrara, F., and Palmisano, G. "A 90-nm CMOS threshold-compensated RF energy harvester," *IEEE J. of Solid-State Circ.*, vol. 46, no. 9, pp. 1985–1997,2011.

[8] Scorcioni, S., Larcher, L., Bertacchini, A., Vincetti, L., and Maini, M. "An integrated RF energy harvester for UHF wireless powering applications," in *2013 IEEE Wireless Power Transfer*, pp. 92–95, 2013.

[9] Theilman, P., Presti, C. C., Kelly, D., and Asbeck, M. "Near zero turn-on voltage high-efficiency UHF RFID rectifier in silicon-on-sapphire CMOS," *IEEE Radio Freq. Integ. Circ. Symp.*, pp. 105–108, 2010.

[10] Kotani K. "Highly efficient CMOS rectifier assisted by symmetric and voltage-boost PV-cell structures for synergistic ambient energy harvesting," in *Proc. IEEE Custom Integ. Circ. Conf.*, pp. 1–4, 2013.

[11] Stoopman, M., Keyrouz, S., Visser, H. J., Philips, K., and Serdijin, W. A. "A self-calibrating RF energy harvester generating 1V at –26.3 dBm," in *Symposium on VLSI Circuits (VLSIC)*, pp. C226–C227, 2013.

[12] Morris, D. H., Avci, U. E., Rios, R., and Young, I. A. "Design of low voltage tunneling-FET logic circuits considering asymmetric conduction characteristics," in *IEEE Journal on Emerging and Selected Topics in Circuits and Systems*, vol. 4, no. 4, pp. 380–388, 2014.

[13] Esseni, D., Guglielmini, M., Kapidani, B., Rollo, T., and Alioto, M. "Tunnel FETs for ultralow voltage digital VLSI circuits: part I—device–circuit interaction and evaluation at device level," in *IEEE Trans. on Very Large Scale Integration (VLSI) Systems*, vol. 22, no. 12, pp. 2488–2498, 2014.

[14] Alioto, M., and Esseni, D. "Tunnel FETs for ultra-low voltage digital VLSI circuits: part II–evaluation at circuit level and design perspectives," in *IEEE Trans. on Very Large Scale Integration (VLSI) Systems*, vol. 22, no. 12, pp. 2499–2512, 2014.

[15] Fulde, M., et al. "Fabrication, optimization and application of complementary Multiple-Gate Tunneling FETs," in *2008 2nd IEEE International Nanoelectronics Conference*, Shanghai, pp. 579–584, 2008.

[16] Paing, T., Falkenstein, E. A., Zane, R., and Popovic, Z. "Custom IC for ultralow power RF energy scavenging," in *IEEE Transactions on Power Electronics*, vol. 26, no. 6, pp. 1620–1626, 2011.

[17] Adami, S.-E., Vollaire, C., Allard, B., Costa, F., Haboubi, W., and Cirio, L. "Ultra-low power autonomous power management system with effective impedance matching for RF energy harvesting," in *2014 8th Int. Conf. on Int. Power Syst. (CIPS)*, pp. 1–6, 2014.

[18] Saini, G., Arrawatia, M., Sarkar, S., and Baghini, M. S. "A battery-less power management circuit for RF energy harvesting with input voltage regulation and synchronous rectification," in *2015 IEEE 58th Int. Mid. Symp. on Circ. and Syst. (MWSCAS)*, pp. 1–4, 2015.

[19] Lanuzza, M., Strangio, S., Crupi, F., Palestri, P., and Esseni, D. "Mixed tunnel-FET/MOSFET level shifters: a new proposal to extend the tunnel-FET application domain," in *IEEE Transactions on Electron Devices*, vol. 62, no. 12, pp. 3973–3979, 2015.

# 8

# TFET-based Power-management Circuit for Nanowatt DC Energy-Harvesting Sources

In this chapter, a Tunnel Field-Effect Transistor (TFET)-based Power-management Circuit (PMC) for DC energy-harvesting (EH) sources is proposed. As explained in the previous chapter, in conventional inductor-based boost converters, the output transistor is shown as an important source of losses when the voltage difference between the drain and the source junctions increases (output voltage and switching node of the boost converter). In order to reduce the reverse losses associated with large reverse bias in the output transistor ($|V_{DS}| > 0.5$ V) an improved boost-converter topology is proposed: two TFET devices in series operate as output transistors, with a voltage applied between them when they are reverse biased. The proposed solution shows improved performance of the inductor-based boost converter at large voltage conversion ratios when compared to the conventional boost topology. A TFET-based startup and controller circuits are designed with a power consumption of 1.2 nW, thus allowing a boost converter operation (50 mV to 500 mV) from a power source delivering 2.5 nW. All the simulated results are based on the Verilog-A *LUT*-based *TFET* models described in Chapter 4.

## 8.1 Motivation

State-of-the-art PMCs have shown that it is possible to extract energy from ultra-low voltage DC sources (sub-0.2 V) such as thermogenerators or Photovoltaic cells [1–4]. The performance of such PMCs with large impedance sources is, however, limited by the minimum amount of power required by the controller circuit. At low-voltage operation, the losses inherent in CMOS-based circuits have to be minimized in order to efficiently extract the low power generated by EH transducers. As TFETs present improved electrical characteristics at low-voltage/power operation, it is of interest to explore this technology in the design of PMCs that not only operate at low voltage (sub-0.1 V) but also at ultra-low power (nW range).

## 8.2 Proposed TFET-based PMC for Ultra-low-power DC Sources

In this section, a TFET-based PMC is proposed for ultra-low-voltage/power DC sources. In Figure 8.1, the top-level architecture of the system is shown. Similar to the previous PMC for RF sources, three different modules are proposed: startup circuit, controller circuit, and boost circuit. The startup module is responsible for providing a power-supply voltage to the boost controller without the use of an external battery while the controller module is responsible for providing control signals to the switches of the boost-converter circuit.

The PMC operation is similar to that explained in the previous chapter, i.e. for maximum power transfer from the EH source to the PMC unit, the boost circuit adapts its input impedance to the impedance of the source.

Once the system starts a synchronous mode of operation [Discontinuous Conduction Mode (DCM)], the output capacitor $C_{OUT}$ is charged to a pre-required value and the load is enabled. When the load is enabled for the first time, the PMC enters a self-sustaining mode (SSM) of operation, i.e. the power capacitors previously charged by the EH source are then directly charged by the output capacitor $C_{OUT}$.

### 8.2.1 Startup Circuit

The principle of operation of the startup circuit shown in Figure 8.1 is the same of that explained in the previous chapter (see Section 7.3.1). As shown in Figure 8.2 and in order to allow a proper startup operation at reduced power levels (nW), changes in the voltage monitor are proposed. The differential-pair of the two-stage comparator (see Figure 8.3(a)) is biased with 30 mV (instead of 50 mV) coming from the voltage reference shown in Figure 8.3(b) [3]. In the voltage reference circuit, the leakage current of M1 is mirrored to the output transistors and the voltage reference is given by the $V_{GS}$ sum of M4, M5, and M6. In contrast to thermionic devices with short channel lengths, the leakage current of TFETs presents a small dependence on the

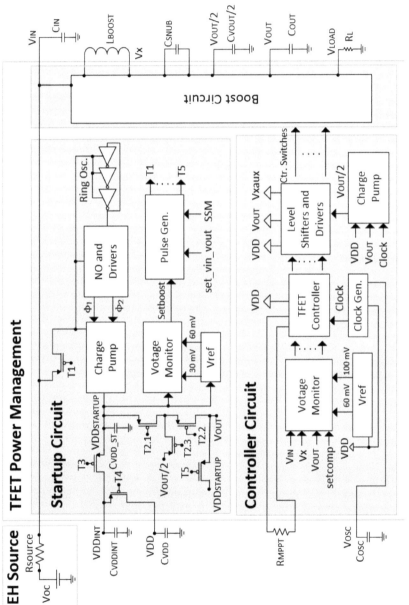

**Figure 8.1**   Top-level architecture of the proposed TFET-based PMC for ultra-low-voltage/power DC sources.

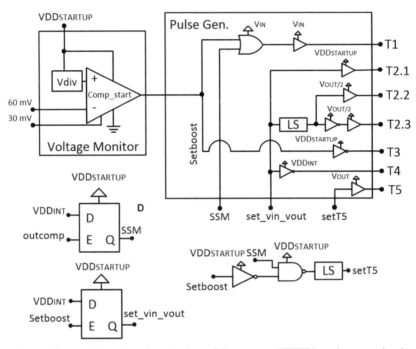

**Figure 8.2**    Digital and Analog circuitry of the proposed TFET-based startup circuit.

drain voltage (at sub-0.6 V) and therefore, the TFET-based voltage reference presents a small dependence on the power supply voltage in the range of 100–600 mV while maintaining ultra-low power consumption.

Prior to the boost conversion operation, the input capacitor $C_{IN}$ is precharged to the $V_{OC}$ value of the EH source and the output capacitor $C_{OUT}$ of the boost converter to the value of $VDD_{STARTUP}$ (200 mV) by the TFET switches controlled by T2.1, T2.2 and T2.3. In the RF PMC presented in the previous chapter, a single p-TFET device is applied between the $VDD_{STARTUP}$ and $V_{OUT}$ nodes. As a consequence, when $V_{OUT}$ increases beyond the value of $VDD_{STARTUP}$, the reverse bias of the *p-TFET* increases and the reverse current degrades the performance of the circuit. For this reason, a possible solution to reduce the reverse losses is to fragment the p-TFET switch in two different TFETs, with a voltage applied between them (for example half the voltage of node $V_{OUT}$) in order to reduce the reverse bias of each one and reverse losses associated. After the pre-charge of the input and output capacitors, a signal *set_vin_vout* is enabled and the switches controlled by T2.1 and T2.2 remain open, with a voltage equal to the voltage node $V_{OUT}/2$ between them.

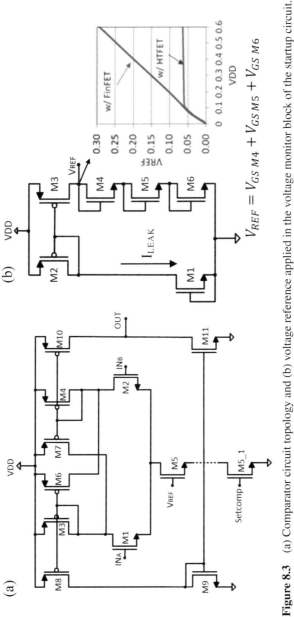

$$V_{REF} = V_{GS\,M4} + V_{GS\,M5} + V_{GS\,M6}$$

**Figure 8.3** (a) Comparator circuit topology and (b) voltage reference applied in the voltage monitor block of the startup circuit.

## 8.2.2 Boost Circuit

The boost converter is required to adapt its input impedance to the impedance of the EH source for maximum power transfer and to increase the output voltage of the source to a required level in order to power a load. In Figure 8.4(a), the TFET-based boost converter topology was shown in the previous chapter to present a good performance for RF EH at μW power levels. As TFETs are designed as reverse-biased *p-i-n* diodes, one of the main challenges is to minimize the reverse current conducted by the output transistor S4 during its reverse-bias state, i.e., when the inductor is being charged and when the boost converter is in idle mode (see Figure 7.11). It was also shown that larger differences between the output and input voltages of the boost converter results in larger reverse-biased TFET S4 (and consequent reverse losses), limiting the voltage operation of the TFET-based circuit.

The previous chapter concluded that increasing the channel width of the heterojunction TFET S4 results in a trade-off between the decrease of its forward losses and increase of reverse losses and consequently, there is an optimum size of S4 that minimizes the conversion losses and increases the boost efficiency for different input power levels. For these reasons, an improved TFET-based boost converter is proposed and shown in Figure 8.4(b). The output transistor S4 is divided in two different ones (S4_1 and S4_2) that are characterized by a voltage applied between them (decreasing their reverse bias) when they operate in the off-state.

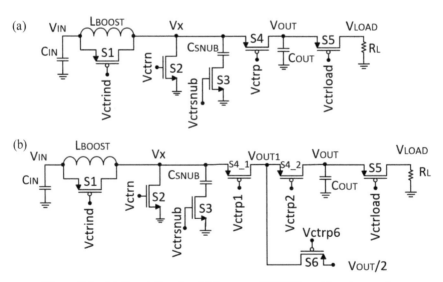

**Figure 8.4**   (a) Conventional and (b) proposed TFET-based boost converter.

The sequence of operation of the proposed boost converter is shown in Figure 8.5(a). During the time interval $\Delta T1$ the input TFET device S2 is closed, and the inductor is charged. The snubber circuit is deactivated, the device S1 is open (off-state), and the voltage at node $Vx$ is approximately 0 V. In order to avoid large reverse losses from the output devices S4_1 and S4_2, the TFET device S6 is closed and a voltage equal to half the voltage of node $V_{OUT}$ is applied to node $V_{OUT1}$. This alleviates the losses of reverse-biased transistors S4_1 and S4_2 by reducing their reverse bias magnitude. The $V_{GS}$ applied to both transistors is 0 V in order to reduce their reverse current.

During $\Delta T2$, the devices S1, S2, S3, and S6 remain in an off-state, and the output transistors S4_1 and S4_2 are closed. The output capacitor is charged by the inductor current with the voltage value of the switching node $Vx$.

During the idle time $\Delta T3$ of the boost converter, the input and output transistors operate in an off-state, with a voltage applied between the two output transistors in order to reduce the conduction of reverse current. In order to attenuate the remaining current in the inductor and avoid large oscillations in the $Vx$ node, the *TFET* device S1 and the snubber circuit are activated.

The boost converter sequence operation is repeated until the voltage at node $V_{OUT}$ reaches a required value, thus enabling an external load RL by closing the device S5 [see Figure 8.4(b)]. The TFET device S5 remains closed until the voltage at node $V_{OUT}$ decreases below a determined threshold point.

In Figure 8.5(b), the sequence of signals applied to the boost converter operating in discontinuous mode are shown. In order to avoid large reverse losses, the boost controller imposes $V_{GS} = 0$ V to all the *TFETs* operating during their off-state (reverse biased).

### 8.2.3 Controller Circuit

The proposed TFET-based controller shown in Figure 8.6 is responsible for providing the control signals applied to the boost converter shown in Figure 8.4(b). The controller is characterized by imposing a $V_{GS} = 0$ V to all the reverse biased TFETs presented in the digital and analog cells, as also the TFET switches presented in the boost converter. This behavior reduces the reverse losses suffered by reverse biased TFETs, thus increasing the PMC efficiency.

An SR latch is responsible for controlling the signals applied to the two output transistors presented in the boost converter. A comparator is required to detect the instant when the inductor current is negative, triggering a *Reset* signal that is applied to the RS latch. Depending on the state of the

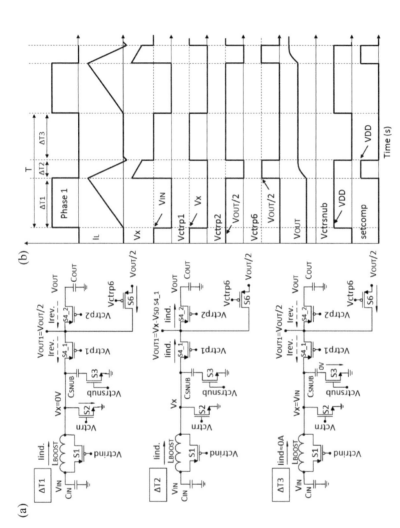

**Figure 8.5** (a) Operation states of the proposed TFET-based boost converter; (b) Operation sequence of the main electrical signals applied to the proposed boost converter.

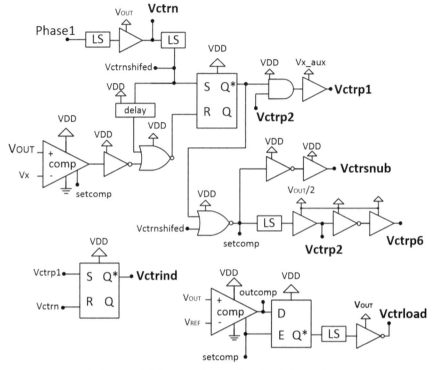

**Figure 8.6** Proposed TFET-based controller circuit for the boost converter.

control signal applied to the input switch S2 (*Vctrn*), the output transistors S4_1 and S4_2 conduct or block current according to the control signals *Vctrp1* and *Vctrp2*. The differential pair of the two-stage comparator is biased with 60 mV coming from the voltage reference of the startup circuit. A second comparator is required to control the device S5 when the output node $V_{OUT}$ reaches a required value, thus activating a load with the control signal *Vctrload*.

When both the input S2 and output S4 devices are operating in the off-state, the control signal *Vctrind* is triggered from an RS latch. In order to maximize the controller efficiency, the two comparators only operate during a small fraction of time, i.e. when the *setcomp* signal is enabled. This signal is enabled during the $\Delta T2$ shown in Figure 8.5, i.e. when the input and output switches of the boost converter are at 0 V.

As explained in the previous chapter, heterojunction TFET-based digital gates are very sensitive to mismatches between digital levels and power supply. Therefore, level shifter (LS) blocks presented in the TFET-based controller are required in order to match the input signals of the digital cells with the power-supply voltage. This method is shown to substantially reduce the power consumption of such cells (see Section 7.3.3).

In order to have a synchronous boost conversion operation, a clock signal is required. The relaxation oscillator shown in Figure 8.7 is responsible for generating a clock signal with a frequency controlled by the capacitor *Cosc*. The $R_{MPPT}$ is responsible for adjusting the duty cycle of the *Phase 1* signal that triggers the *Vctrn* signal applied to the input transistor S2 of the boost converter.

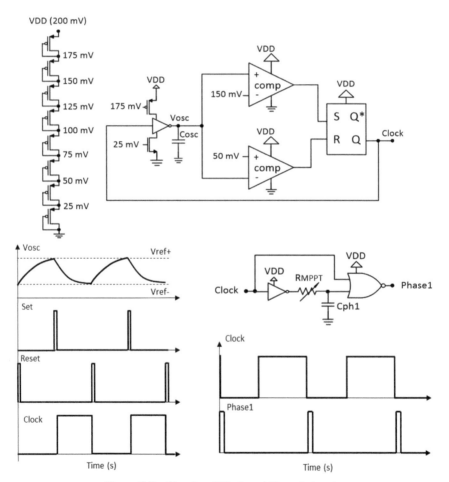

**Figure 8.7**    Circuits of Clock and Phase 1 signals.

The $V_{OUT}/2$ source is generated by a voltage divider charge pump as shown in Figure 8.8. The proposed circuit for TFET operation requires two non-overlapped clock signals generated by a non-overlapped *NO* circuit powered by $V_{OUT}$. In order to improve the conversion efficiency, the reverse-biased TFETs during each region of operation are characterized by a $V_{GS} = 0$ V (see Table 8.1).

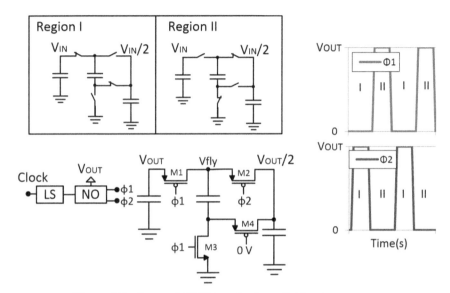

**Figure 8.8** Proposed TFET-based voltage divider charge pump.

**Table 8.1** Bias condition of TFETs applied to the voltage divider CP in steady-state conditions

|        |       | I |       | II |
| --- | --- | --- | --- | --- |
| Region | State | Bias | State | Bias |
| M1 (p) | on | $V_{GS} = -V_{OUT}$ <br> $V_{DS} < 0$ V | off | $V_{GS} \approx 0$ V <br> $V_{DS} < 0$ V |
| M2 (p) | off | $V_{GS} \approx 0$ V <br> $V_{DS} < 0$ V | on | $V_{GS} = -V_{OUT}/2$ <br> $V_{DS} < 0$ V |
| M3 (n) | off | $V_{GS} \approx 0$ V <br> $V_{DS} > 0$ V | on | $V_{GS} = V_{OUT}$ <br> $V_{DS} > 0$ V |
| M4 (p) | on | $V_{GS} = -V_{OUT}/2$ <br> $V_{DS} < 0$ V | off | $V_{GS} \approx 0$ V <br> $V_{DS} > 0$ V |

## 8.3 Simulated Results

This section presents the simulated results of the TFET-based PMC circuit shown in Figure 8.1 for ultra-low-power EH sources. The PMC is designed with heterojunction III–V TFET models (InAs-GaSb, $L_G$ = 40 nm) as described in Chapter 4. In order to seek high efficiencies at nW power levels, the boost converter operates in DCM.

For maximum power extraction from an EH source, the input impedance of the boost converter should equal the impedance of the source. In an ideal boost converter, the input impedance can be approximated as expressed by (8.1):

$$Z_{IN} = \frac{2L}{t_1^2 f_s} \cdot \left(1 + \frac{V_{IN}}{V_{OUT} - V_{IN}}\right)^{-1} \tag{8.1}$$

$$i_L = \frac{V_{IN} \cdot t_1}{L} \tag{8.2}$$

In this work, the EH source is simulated with two different impedances, 1 M$\Omega$ and 100 k$\Omega$. When considering a fixed inductor $L$, fixed boost frequency $f_S$, and $V_{OUT} \gg V_{IN}$, the input impedance of the boost converter can be controlled by $t_1$, i.e. the on-time of the input transistor S2. In this work, an inductor with 470 µH and a boost frequency of 100 Hz are considered. As expressed by (8.2), the inductor current and the inductance value are inversely proportional. Therefore, in order to avoid large forward losses in the switches presented in the boost converter, a large inductor size is preferred. The low operation frequency allows for the reduction of switching losses presented in the controller circuit.

As shown in Figure 8.9(a), a *Cosc* value of 7 pF presented in the clock circuit generates a clock frequency of 100 Hz. In Figure 8.9(b), the phase 1 on-time $t_1$ required for different input power levels is presented. For source impedances ($R_{SOURCE}$) with 1 M$\Omega$ and 100 k$\Omega$, $R_{MPPT}$ values (see Figure 8.7) of respectively 3.8 M$\Omega$ and 14 M$\Omega$ are shown to be adequate for maximum power point tracking (MPPT) in the $V_{IN}-V_{OUT}$ range considered: $V_{IN}$ between 0.05–0.2 V and $V_{OUT}$ between 0.5–0.7 V.

In Figure 8.10, the performance of the conventional and proposed *TFET*-based boost converters shown in Figure 8.4 is compared considering a DC EH source with 1 M$\Omega$ and different boost converter input voltages. A load of 6.25 M$\Omega$, 25 M$\Omega$, and 100 M$\Omega$ is enabled (for input power levels of respectively 40 nW, 10 nW and 2.5 nW) when the output voltage of the boost converter reaches a threshold value of 515 mV. It is shown that in the conventional topology, there is an adequate output transistor size S4 that minimizes

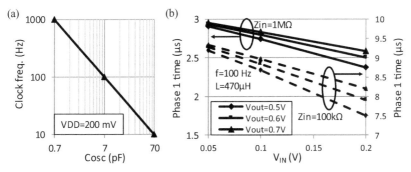

**Figure 8.9** (a) Clock frequency in function of Cosc; (b) Phase 1 time required for MPPT. L = 470 μH, f = 100 Hz.

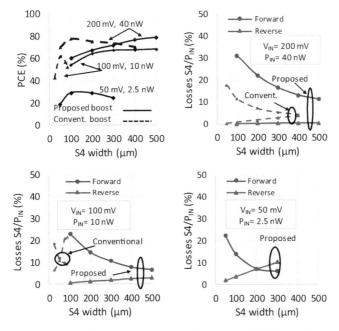

**Figure 8.10** Performance of the conventional and proposed TFET-based boost converters considering a DC energy harvesting source with 1MΩ. $L_{BOOST}$ = 470 μH, WS1 = 5 μm, WS2 = 1mm, WS3 = 10 μm, WS5 = 50 μm, WS6 = 200 μm.

the conduction of reverse current (when reverse biased) and forward losses. In contrast, the proposed TFET-based boost converter allows for the reduction of forward losses with larger S4 sizes (S4_1 + S4_2), maintaining low reverse losses.

As an example, the performance of the conventional boost converter with an input voltage of 0.1 V and output voltage of 0.5 V is degraded due to the large reverse losses suffered by the output transistor S4 when reverse biased ($V_{DS} = 0.5$ V during $\Delta$T1 and $V_{DS} = 0.4$ V during $\Delta$T3). In contrast, the proposed converter characterizes S4_1 and S4_2 with a reverse bias of $V_{DS} = 0.25$ V during $\Delta$T1 and S4_1 (S4_2) with $V_{DS} = 0.15$ V ($V_{DS} = 0.25$ V) during $\Delta$T3, thus reducing the conduction of reverse current.

The combination of sub-nW power consumption of the TFET-based startup (614 pW) and controller circuits (580 pW) shown in Figure 8.11 and the decrease of reverse current conduction in output transistors S4_1 and S4_2 allow the proposed boost converter to operate with input power levels as low as 2.5 nW and PMC with 29% of PCE ($V_{IN} = 50$ mV, $V_{OUT} = 0.5$ V).

In Figure 8.12 the performance of the proposed TFET-based boost converter is shown considering an output transistor S4 (S4_1 + S4_2) width of 200 $\mu$m. For an input power of 10 nW ($V_{IN} = 0.1$ V, $R_{SOURCE} = 1$ M$\Omega$) the boost converter is simulated with $C_{IN} = C_{OUT} = 0.1$ $\mu$F whereas for an input power of 100 nW ($V_{IN} = 0.1$ V, $R_{SOURCE} = 100$ k$\Omega$) the capacitance values of $C_{IN}$ and $C_{OUT}$ are simulated as 1 $\mu$F. The results show that the proposed circuit is able to increase a low input voltage value of 0.1 V to 0.7 V.

The distribution of power losses in the boost converter is presented in Figure 8.13 ($I_{LOAD} = 100$ nA for $P_{IN} = 10$ nW and $I_{LOAD} = 1\mu$A for $P_{IN} = 100$ nW). One can observe that larger output voltage values result

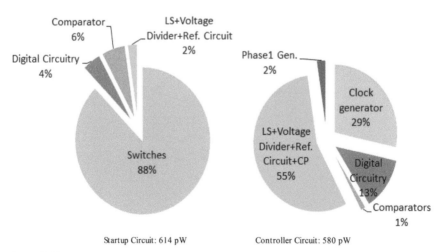

**Figure 8.11**   Distribution of power losses in the proposed TFET-based startup and controller circuits.

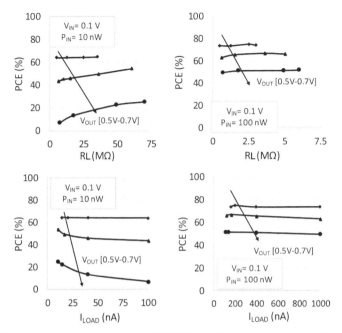

**Figure 8.12** Performance of the proposed TFET-based boost converter for different voltage conversion ratios. WS4 = 200 μm (WS4_1 = WS4_2 = 100 μm).

in larger losses in the TFET switches S1 and S5 when the input power is low. When the load is not enabled, the increase of $|V_{DS}|$ in S5 at larger output voltage values results in an increase of leakage current and consequent power losses. The switch S1 is also shown as an important source of power losses. During $\Delta T2$, the increase of voltage at node $Vx$ with larger output voltage values imposes a high reverse bias on this TFET device, thus increasing its reverse losses.

In Figure 8.14, the load on-time for different input power levels and output voltages is presented. It is shown that the TFET-PMC powered by an EH source delivering 10 nW ($V_{IN} = 0.1$ V, $C_{IN} = C_{OUT} = 0.1$ μF) can enable a load with 100 nA ($V_{OUT} = 0.7$ V) during 20 ms, i.e. two conversion cycles of 10 ms. A similar value is achieved considering a load with 1 μA and a source of 100 nW ($V_{IN} = 0.1$ V, $C_{IN} = C_{OUT} = 1$ μF).

In Figure 8.15 the transient simulation of the TFET-based PMC is presented, considering an EH source with an open circuit voltage of 200 mV and 1 MΩ. With MPPT, the input impedance of the boost converter equals the impedance of the source and an input voltage of 0.1 V ($P_{IN} = 10$ nW)

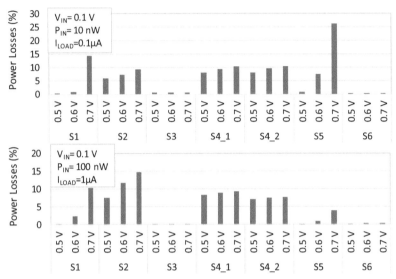

**Figure 8.13**    Distribution of power losses in the proposed boost-converter for different voltage conversion ratios ($V_{IN}$ = 0.1 V, $V_{OUT}$ = 0.5 V, 0.6 V and 0.7 V) considering an output load of 100 nA ($P_{IN}$ = 10 nW) and 1 µA (for $P_{IN}$ = 100 nW).

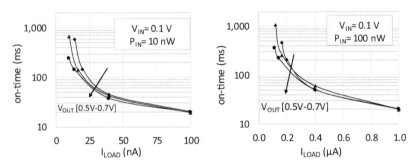

**Figure 8.14**    Load on-time for different input power levels and output voltage.

is observed. One can observe that prior to the boost conversion operation the input $C_{IN}$ and output $C_{OUT}$ capacitors of the boost converter are pre-charged to the open circuit voltage of the source and $VDD_{STARTUP}$ respectively. Once charged, the power-supply node of the controller is enabled (VDD) and the boost converter starts a synchronous mode of operation. When the capacitor at the output voltage node $V_{OUT}$ is charged beyond a threshold voltage value of 515 mV, a load is enabled until the capacitor $C_{OUT}$ discharges below a threshold voltage of 500 mV. When the load is enabled for the first time,

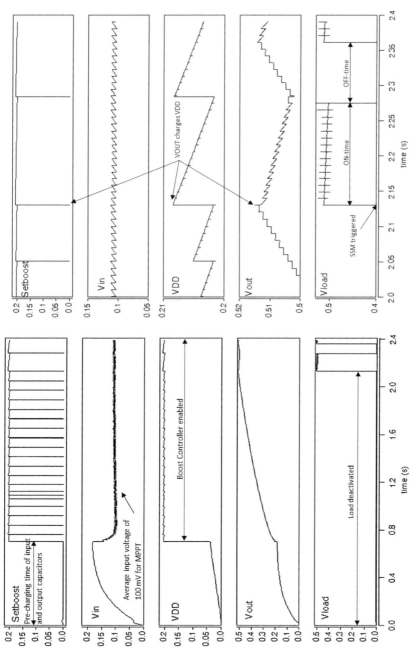

**Figure 8.15** TFET-based PMC transient behavior considering $P_{IN} = 10$ nW, $V_{IN} = 0.1$ V, $V_{OUT} = 0.5$ V, RL = 25 MΩ, L = 470 μH.

**Table 8.2**　Performance comparison with power management circuits from the literature

| Ref. | [1] | [2] | [3] | [4] | This Work |
|---|---|---|---|---|---|
| Tech. | 130 nm CMOS | 65 nm CMOS | 180 nm CMOS | 320 nm CMOS | 40 nm TFET |
| Year | 2010 | 2013 | 2016 | 2016 | 2017 |
| Battery for Start-up | Yes (ext. capacitor) | No | No | No | No |
| $V_{OUT}$ | 1 V | 1.2 V | 3 V | 1 V | 0.5–0.7 V |
| PCE | 37%, $V_{IN}$ = 50 mV | | 40%, $V_{IN}$ = 360 mV | | 29%, 50–500 mV |
| | $P_{IN}$ = 370 nW 68%, $V_{IN}$ = 100 mV $P_{IN}$ = 6.8 µW | 70%, $V_{IN}$ = 50 mV $P_{IN}$ = 403 µW | $P_{IN}$ = 10 nW 75%, $V_{IN}$ = 360 mV $P_{IN}$ = 100 nW | 50%, $V_{IN}$ = 124 mV, $P_{IN}$ = 10.5 µW | $P_{IN}$ = 2.5 nW 50%, 100–700 mV $P_{IN}$ = 100 nW |
| Energy Source | Thermal | Thermal | Solar | Solar | DC energy harvesting |

the circuit enters an SSM of operation, i.e. the output capacitor is responsible for charging the power sources of the startup circuit and controller.

In Table 8.2, a comparison between the performance of the proposed TFET-based PMC and recent power management units from the literature is presented. It is shown that the inclusion of III–V heterojunction TFETs in PMCs shows promising results for the EH field at power levels in the nW range.

## 8.4 Impact of TFET-based Circuit Layout and Parasitics

In the results presented in this chapter, no layout parasitics were included in the PMC performance analysis. As opposite to conventional thermionic devices, the different doping structure in TFETs requires changes in the cell layouts due to the non-sharing possibility between contacts. In order to analyze the impact of parasitics in the circuit performance, further work in the device structure and layout is required. Vertical TFET structures are under investigation to reduce the device footprint area and consequent circuit overhead compared to thermionic devices, and also due to the feasibility of the heterojunction structure implementation [5–7].

At nW power levels, the layout and associated resistance parasitics are expected to have some impact on the performance of the proposed PMC and therefore, improved models are required for a proper circuit analysis. Although the presented results do not include pad connection losses and parasitics, the good performance shown by the proposed TFET-based boost converter demonstrates the potential of using TFETs in ultra-low power conversion for EH applications.

## 8.5 Chapter Summary

In this chapter, solutions are proposed to increase the range of inductor-based boost converter operation by reducing the reverse losses of the TFET output transistor. With the proposed techniques, the output voltage of the boost converter can be extended to 0.7 V from an input voltage of 0.1 V. The results show that TFETs can enable the extraction from DC EH sources that not only present very low voltage levels (sub-0.1 V) but also very low power levels (a few nW). It is shown by simulations that the proposed TFET-based PMC (designed with III-V heterojunction TFETs) can sustain itself from a 2.5 nW source, powering a load (0.5 V) from an input voltage of 50 mV with 29% of efficiency.

## References

[1] Carlson, E. J., Strunz, K., and Otis, B. P. "A 20 mV input boost converter with efficient digital control for thermoelectric energy harvesting," in *IEEE JSSC*, vol. 45, no. 4, pp. 741–750, 2010.

[2] Weng, P. S., Tang, H. Y., Ku, P. C., and Lu, L. H. "50 mV-input batteryless boost converter for thermal energy harvesting," in *IEEE TCAS I: Regular Papers,* vol. 48, no. 4, pp. 1031–1041, 2013.

[3] El-Damak, D., and Chandrakasan, A. P. "A 10 nW–1 μW power management IC With integrated battery management and self-startup for energy harvesting applications," in *IEEE Journal of Solid-State Circuits*, vol. 51, no. 4, pp. 943–954, 2016.

[4] Dini, M., Romani, A., Filippi, M. and Tartagni, M. "A nanocurrent power management IC for low-voltage energy harvesting sources," *IEEE TPE,* vol. 31, no. 6, pp. 4292–4304, 2016.

[5] Avci, U. E., Morris, D. H., and Young, I. A. "Tunnel field-effect transistors: prospects and challenges," in *IEEE Journal of the Electron Devices Society*, vol. 3, no. 3, pp. 88–95, 2015.

[6] Kim, M. S., Cane-Wissing, W., Li, X., Sampson, J., Datta, S., Gupta, S. K., et al., "Comparative area and parasitics analysis in FinFET and heterojunction vertical TFET standard cells," *ACM J. Emerg. Tech. Comp. Syst.* 12:38, 2016.

[7] Kim, M. S., Cane-Wissing, W., Sampson, J., Datta, S., Narayanan, V., Gupta, S. K. "Comparing energy, area, delay tradeoffs in going vertical with CMOS and asymmetric HTFETs," in *2015 IEEE Computer Society Annual Symposium on VLSI*, pp. 303–308, 2015.

# 9

# Final Conclusions

This book presents several key points on the design of ultra-low power circuits based on tunnel field-effect transistors (TFETs) for energy-harvesting (EH) applications. At a device level, the Band-to-Band Tunneling (BTBT) carrier injection mechanism based on the Zener tunneling effect characterizes TFETs with an inverse sub-threshold slope (SS) below the limited 60 mV/dec at room temperature of conventional thermal-dependent switching devices. This characteristic allows for the reduction of the operating voltage of circuits, without jeopardizing the leakage current of the device and consequent static power of circuits. This is presented as the main advantage of TFETs when compared to conventional complementary metal-oxide-semiconductor (CMOS) devices.

The doping concentration and profile in the source/drain regions directly influences the performance of TFETs. The increase of source-doping concentration allows for an increased electrical field magnitude applied between the source-channel regions and therefore increased BTBT current with a consequent increase of the leakage current and SS. Equal source- and drain-doping concentrations results in TFETs with ambipolarity, i.e., reverse BTBT occurs at the drain-channel interface. Therefore, a drain-doping concentration lower than the source-doping counterpart is required to reduce either the leakage current or the reverse BTBT current. As the BTBT current is highly dependent on the electric field applied between the source-channel regions, a uniform source-doping profile is preferred when compared to a Gaussian one.

Compared to silicon-based TFETs, the use of lower energy band gap materials (Ge or InAs) is shown to improve the device performance at lower gate voltage magnitudes with a consequent degradation of the leakage current. This behavior is directly related to the decrease of barrier width between the source-channel regions and increase of BTBT probability. Several works have shown by experiments that TFETs designed with III–V materials can achieve larger drive currents in comparison to Si-TFETs but still lower than the on-currents observed by conventional CMOS devices. In addition, further improvements in the development of defect-free III–V materials are required in order to maintain the low leakage current of III–V TFETs without degrading the SS of the device.

As TFETs are designed as reverse-biased *p-i-n* diodes, there is a need to change conventional circuit topologies that characterize transistors with forward- and reverse-bias states. This is the case of gate cross-coupled charge pumps and gate cross-coupled rectifiers. At reverse bias, the intrinsic *p-i-n* diode of TFETs is forward-biased and the reverse current becomes important, thus degrading the conversion efficiency of such front-end circuits. A solution to attenuate the reverse current of TFETs under reverse-bias conditions is to decrease the gate voltage magnitude (in relation to source). With the proposed TFET-based charge pump and TFET-based rectifier, this solution is shown by simulations (using heterojunction TFET models from the literature) to extend the voltage/power range of operation in comparison to conventional topologies.

- The proposed charge-pump is shown to present a similar performance to the conventional topology for input voltages between 160 mV and 400 mV, and larger efficiencies at larger inputs. Although increased switching losses are caused by the auxiliary circuit, the improved efficiency of the proposed charge pump is due to the reduction of the reverse losses suffered by the main transistors inside the stage when subjected to large reverse bias;
- The proposed rectifier presents an extended RF-voltage operation when compared to the conventional topology when considering a frequency of operation of 100 MHz. In contrast, at large frequency (e.g., 915 MHz) the auxiliary circuitry of the proposed rectifier is shown to produce important switching losses in the stage, degrading the efficiency at RF input power levels in the range of –25 dBm and –40 dBm (corresponding to RF $V_{AC}$ between 0.2 V and 0.6 V). Despite the increased switching losses at large frequencies, the proposed rectifier performs better than the conventional one at large RF $V_{AC}$ (>0.6 V).

Two power-management circuits (PMC) were designed based on the heterojunction TFET models from the literature (InAs-GaSb, $L_G$ = 40 nm). One is designed to interface RF energy harvesting sources and the second to interface DC sources. Both PMCs present a startup, a controller, and boost-converter modules. In the first PMC, the challenges of using TFETs in inductor-based boost converters are identified. If the output transistor of the boost converter is largely reverse biased, i.e., the difference between the output voltage and input voltage of the boost converter is large, then large reverse current degrades the boost converter performance. This characteristic limits the performance of TFET-based boost converters to low voltage operation. Nevertheless, the proposed TFET-PMC shows promising results at available RF power levels below –20 dBm (f = 915 MHz). For an available power of –25 dBm, the proposed boost converter is able to deliver 1.1 μW of average power to a load (0.5 V) with a boost efficiency of 86 %.

With the challenges of using TFETs in inductor-based boost converters identified, a different boost converter topology is proposed: two TFET devices in series operate as output transistors, with a voltage applied between them when they are reverse biased. The proposed solution shows improved performance of the inductor-based boost converter at large voltage conversion ratios when compared to the previous boost topology of the RF-PMC. The output voltage of the boost converter can be extended to 0.7 V from an input voltage of 0.1 V (previous output was 0.5 V). In order to evaluate the performance of the proposed PMC with the proposed inductor-based boost converter, a DC source with fixed impedance was simulated as the input source. The results show that TFETs can enable the extraction from DC energy harvesting sources that not only present very low voltage levels (sub-0.1 V) but also very low power levels (a few nW).

## 9.1 Summary of Book Contributions

The main contributions of this book are listed as follows:

- At a device level, explore the dependence of the electrical properties of TFETs on several physical parameters such as: doping concentration and doping profile in source/drain regions, and dielectric permittivity and EOT of gate oxides and materials (Si, Ge, and InGaAs). This task identified key parameters for optimized TFETs to be applied in ultra-low power circuits under forward and reverse bias conditions;
- Explore the performance of TFETs in analog and digital design by extracting key figures of merit (FOM). The FOM were compared to those of Si-FinFET devices ($L_G$ = 20 nm), thus allowing to identify the voltage range where TFETs present superior electrical performance. This task was performed with Verilog-A-based look-up table models from the literature that describe the behavior of homojunction TFETs (InAs, $L_G$ = 20 nm) and heterojunction TFETs (InAs-GaSb, $L_G$ = 40 nm);

- Explore the performance of TFETs in front-end charge pumps and rectifiers. The limitations of using TFETs in conventional front-end topologies were identified, with the proposal of solutions at circuit level that increase the voltage/power range operation of such circuits;
- Design of TFET-based PMC for RF EH applications ($\mu$W). Circuit techniques to improve the efficiency of PMCs designed with TFETs are proposed for increased efficiency. Changes in conventional inductor-based boost converters are proposed in order to overcome the lack of body diodes in TFETs. The limitation of using TFETs in inductor-based boost converters are identified;
- A PMC for DC energy harvesting sources (nW) is proposed. With the limitations of using TFETs in inductor-based boost converters identified, changes in the converter are proposed. The results show that when compared to the use of conventional inductor-based boost converters, the proposed solution can increase the voltage operation by increasing the output voltage of the converter and consequent voltage gain.

## 9.2 Future Work

The following points show several promising tasks to further extend the state of the art of TFET-based circuit design for energy harvesting applications:

- An improved TFET compact model is required to further evaluate the performance of the proposed circuits. The impact of process and temperature variations, electrical noise and layout parasitics has to be considered in order to evaluate the integrity of TFETs at ultra-low power levels. An analytical model describing the dynamic and static behavior of the TFET under all regions of operation for both $n$ and p-type configurations would speed up the simulation results compared to Verilog-A look-up tables;
- Experimental validation of the proposed circuits is required in order to validate the TFET-based circuit design techniques proposed in this book. In order to accomplish this task, several effects that degrade the leakage current of heterojunction III–V TFETs have to be overcome in order to experimentally demonstrate the performance shown by simulations at a device level. Therefore, further investigation in III–V and novel materials that currently present high bulk and interface defects is required in order to achieve III–V based-TFETs with the maturity level of silicon-based devices.

# Glossary

| | |
|---|---|
| $\lambda_o$ | Wavelength of transmitted signal |
| $\triangle\Phi$ | Screening length |
| AC | Alternate current |
| Al | Aluminum |
| $Al_2O_3$ | Aluminum oxide |
| BTBT | Band-to-Band Tunneling |
| CC | Coupling capacitor |
| $C_{DEP}$ | Depletion capacitance |
| $C_{GD}$ | Gate-to-drain capacitance |
| $C_{GS}$ | Gate-to-source capacitance |
| $C_{INT}$ | Interface capacitance |
| $C_L$ | Load capacitance |
| CMOS | Complementary metal-oxide-semiconductor |
| CNT | Carbon nanotube |
| $C_{OX}$ | Oxide capacitance |
| DC | Direct current |
| DCM | Discontinuous conduction mode |
| DG | Double-gate |
| DOS | Density of states |
| E | Energy |
| $E_F$ | Energy Fermi level |
| $E_G$ | Energy banggap |
| EH | Energy harvesting |
| EOT | Equivalent oxide thickness |
| F | Electric field |
| $F_c$ | Fermi Dirac distribution of conduction band |
| FDSOI | Fully depleted silicon on insulator |
| $F_H$ | Fin height |
| $f_{MAX}$ | Maximum oscillation frequency |
| FOM | Figure of merit |
| $f_s$ | Boost frequency |

| | |
|---|---|
| $f_T$ | Cut-off frequency |
| $F_V$ | Fermi Dirac distribution of valence band |
| $F_W$ | Fin Width |
| GAA | Gate all around |
| GaAs | Gallium arsenide |
| $G_{BTBT}$ | Band-to-Band tunneling generation rate |
| GBW | Gain bandwidth product |
| GCCCP | Gate cross-coupled charge pump |
| GCCR | Gate Cross-coupled rectifier |
| Ge | Germanium |
| GeOI | Germanium on insulator |
| GFETs | Graphene field effect transistor |
| $g_m$ | Transconductance |
| $g_m/I_{DS}$ | Transconductance efficiency |
| $G_R$ | Receiver antenna gain |
| $G_T$ | Transmitter antenna gain |
| $HfO_2$ | Hafnium oxide |
| HP | High performance |
| HTFET | Heterojunction TFET |
| IC | Integrated circuit |
| $I_E$ | Esaki tunneling current |
| InAs | Indium arsenide |
| InGaAs | Indium gallium arsenide |
| IoT | Internet of things |
| ISM | Industry-science-medical |
| $I_Z$ | Zener tunneling current |
| k (x) | Quantum wave vector |
| $L_G$ | Gate length |
| LP | Low performance |
| LS | Level shifter |
| LSTP | Low standby power |
| LUT | Look-up table |
| m* | Effective mass of the carrier |
| MBE | Molecular beam epitaxy |
| MM | More Moore |
| $MoS_2$ | Molybdenum disulfide |
| MOSFET | Metal-oxide-semiconductor field effect transistor |
| MPPT | Maximum power point tracking |
| MtM | More than Moore |

| | |
|---|---|
| $N_A$ | Source doping concentration |
| $N_C$ | Density of states in conduction band |
| $N_D$ | Drain doping concentration |
| NDR | Negative differential resistance |
| $N_{FIN}$ | Number of FINS |
| NMOS | N-type MOSFET |
| NO | Non-overlapped |
| $N_V$ | Density of states in valence band |
| NW | Nanowire |
| PCE | Power conversion efficiency |
| PMC | Power-management circuit |
| PMOS | P-type MOSFET |
| $P_R$ | Received power |
| $P_T$ | Transmitted power |
| PTM | Predictive technology model |
| PV | Photovoltaic |
| R | Propagation distance between transmitter and receiver |
| $R_A$ | Real part of antenna impedance |
| $R_D$ | Drain resistor |
| RF | Radio-frequency |
| $R_G$ | Gate resistor |
| RL | Output load |
| $R_S$ | Source resistor |
| SCTJ | Surface channel tunnel junction |
| SG | Single gate |
| Si | Silicon |
| SiGe | Silicon-Germanium |
| $SiO_2$ | Silicon-oxide |
| SOI | Silicon on insulator |
| SPICE | Simulation program with integrated circuit emphasis |
| SS | Inverse sub-threshold slope |
| SSM | Self-sustaining mode |
| STT | Surface tunnel transistor |
| TAT | Trap-assisted tunneling |
| $T_{BTBT}$ | Tunneling transmission probability |
| TCAD | Technology computer-aided design |
| $t_{CH}$ | Channel thickness |
| TEG | Thermo-electric generator |
| TFET | Tunnel field-effect transistor |

| | |
|---|---|
| TiN | Titanium nitride |
| $t_{OX}$ | Oxide thickness |
| U | Barrier potential |
| UHF | Ultra-high frequency |
| UTB | Ultra thin body |
| $V_A$ | Peak amplitude of the antenna |
| $V_{DD}$ | Power supply voltage |
| $V_{DS}$ | Drain-to-source voltage |
| $V_{GS}$ | Gate-to-source voltage |
| VLSI | Very large scale integration |
| VOC | Open circuit voltage |
| $V_{TH}$ | Threshold voltage |
| $V_{THD}$ | Drain-threshold voltage |
| $W_{FIN}$ | FinFET channel width |
| WKB | Wentzel-Kramers-Brillouin |
| $Z_rO_2$ | Zirconium dioxide |
| $\alpha$ | Switching activity factor |
| $\varepsilon_r$ | Relative permittivity or dielectric constant |
| $\lambda$ | Natural length |
| $\Psi$ | Wave function of carrier |
| $\hbar$ | Reduced Planck constant |

# Index

# About the Authors

**David Cavalheiro** received the M.Sc. degree in electrical engineering from the New University of Lisbon, Portugal, in 2012 and the Ph.D. degree in electronic engineering from the Universitat Politècnica de Catalunya, Spain, in 2017. His current research activities include the design of ultra-low power circuits for conversion applications. Other interests are power management circuits, emerging transistor technologies and energy harvesting.

**Francesc Moll** received the equivalent of the M.Sc. degree in physics from the University of Balearic Islands (UIB), Spain, in 1991, and the Ph.D. degree in electronic engineering from the Technical University of Catalonia (UPC), Spain, in 1995. Since 1991, he has been with the Department of Electronic Engineering, Technical University of Catalonia, where he is currently an Associate Professor. His current research activities include methods for energy harvesting and low-power integrated circuit design.

**Stanimir Valtchev** received the M.Sc. from TU Sofia in 1974, and the PhD degree from IST in Lisbon. He worked on semiconductor technology, medical equipment, and then as a researcher in industrial electronics (laser supplies and high-frequency power converters). He is currently Auxiliary Professor in UNL and Invited full professor in BFU, Bulgaria. His research includes power converters (also resonant and multilevel), energy harvesting, wireless energy transfer, electric vehicles, energy management and storage, bio energy-harvesting, and biosensors.